NIST GCR 03-853

U.S. Department of Commerce
Technology Administration
National Institute of Standards and Technology

Office of Applied Economics
Building and Fire Research Laboratory
Gaithersburg, MD 20899

# BridgeLCC 2.0 Users Manual
Life-Cycle Costing Software for the Preliminary Design of Bridges

Mark A. Ehlen

Prepared For:
    Amy S. Rushing
    Office of Applied Economics
    Building and Fire Research Laboratory
    National Institute of Standards and Technology
    Gaithersburg, MD 20899-8603

Under Contract NA1341-02-W-0884


September 2003

**U.S. DEPARTMENT OF COMMERCE**
    Donald L. Evans, Secretary

**TECHNOLOGY ADMINISTRATION**
    Philip J. Bond, Under Secretary for Technology

**NATIONAL INSTITUTE OF STANDARDS AND TECHNOLOGY**
    Arden L. Bement, Jr., Director

# Abstract

BridgeLCC 2.0 is user-friendly software developed by the National Institute of Standards and Technology to help bridge designers determine the cost effectiveness of alternative bridge designs, construction and repair strategies, and construction materials. The software uses a life-cycle costing methodology based on the ASTM standard practice for life-cycle costing and a cost classification scheme developed by NIST. This user manual describes the functions and settings in BridgeLCC and includes example analyses that illustrate its use.

BridgeLCC 2.0 is designed to run on Windows® 95, 98, 2000, NT, and XP. Although the software is specifically tailored to highway bridges, it can also be applied to pavements, piers, and other civil infrastructure.

## Keywords

bridge design; building economics; construction; costs; economic analysis; engineering economics; life-cycle costing; maintenance costs; operations costs; uncertainty and risk analysis; value engineering

## Acknowledgments

Thanks are given to Muthial Kasi and Elissa Schneider of Alfred Benesch & Co., Tom Canick of FHWA, and Amy Rushing, Robert Chapman, Christine Izzo, Chi Leng, and Harold Marshall of NIST for assistance and feedback during the programming of BridgeLCC 2.0.

# Table of Contents

## List of Figures

**List of Tables**

# 1. Introduction

## 1.1 Background

The Building and Fire Research Laboratory, one of seven laboratories at the National Institute of Standards and Technology, performs research in diverse areas such as structural design, new-technology construction materials, automated construction techniques, and fire resistance of building systems. Its staff work to develop and provide to industry new cost-effective materials, designs, and processes for buildings, bridges, and other structures in built environments.

In support of these objectives, the Office of Applied Economics develops and provides industry with tools that determine the cost effectiveness of building-related alternatives. In the case of designing, building, and maintaining a highway bridge, the cost-effective bridge designs, construction processes, and repair strategies are those that minimize the costs to the owners and users of the bridge over its life or *life cycle*. For state and local agencies that maintain many bridges of different ages and uses, reducing these bridge life-cycle costs reduces the aggregate cost of providing their regions' transportation infrastructure and thereby the tax burden on its citizens.

Bridge engineers designing a new bridge or repairing an existing bridge will typically – and are often required to – compare and choose from several alternative strategies, such as "steel structure vs. concrete structure" or "repair the structure vs. replace the structure." In many cases the engineer has an existing, "base case" technique or strategy, and "alternatives" that represent specific changes to this base case. Currently, for alternatives that provide the same technical performance, including code compliance and service life, construction costs are typically used to compare and ultimately decide on the design strategy. But an alternative with higher initial construction costs may have significantly lower operation, maintenance, and repair costs, and therefore life-cycle costs. Life-cycle cost analysis allows the engineer to determine which alternative is cost effective over its intended life.

BridgeLCC is user-friendly, Windows® software specifically designed to help engineers, material specialists, and budget analysts determine the life-cycle cost effectiveness of their bridge designs and processes. The user defines his or her project (such as building a bridge), defines the alternatives (such as making the bridge with steel versus making the bridge with concrete), and then compiles the costs of building, maintaining, and then disposing of each of these alternatives. Costs include project costs incurred by the agency responsible for the structure (*agency costs*), costs incurred by drivers on the highway that are inconvenienced by bridge construction and other bridge activity (*user costs*), and costs incurred by third parties who are not direct users of the structure but are impacted by construction and repair activity (*third-party costs*).

Once the costs are compiled, the user compares the life-cycle costs of the alternative bridges or processes. The alternative with the lowest life-cycle cost, all other factors being equal, is the cost-effective bridge. The user utilizes the cost classification in BridgeLCC to compare the technical advantages and disadvantages of each alternative in life-cycle cost terms.

BridgeLCC uses a life-cycle costing methodology based on the ASTM practice for measuring the life-cycle costs of buildings and building systems (ASTM E 917) and a NIST cost classification scheme for comparing life-cycle costs of alternatives. The ASTM practice insures that the cost calculations follow accepted practice; the scheme helps the user account for all project costs, properly categorize them, and then compare breakdowns of the alternatives' life-cycle costs.

Figure 1 illustrates how BridgeLCC provides the framework for following the ASTM practice and for categorizing and comparing costs. The **Cost Summary** window serves as a "home page," where life-cycle cost totals are displayed, alternatives' costs can be accessed, and a step-wise list can be used to access the most common tasks.

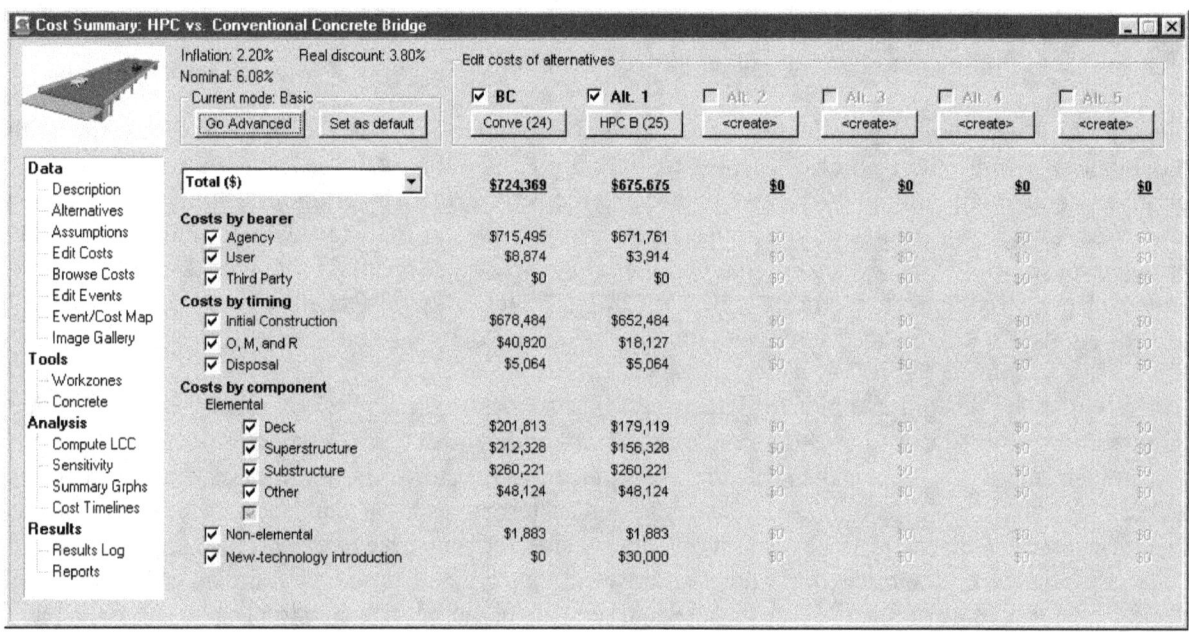

**Figure 1. Cost Summary Window**

Most of the ASTM-consistent steps required to complete a life-cycle cost analysis can be accessed under **Data**, **Tools**, **Analysis**, and **Results** in the **Cost Summary** window, including how to

- describe the overall project and the alternatives under consideration;
- make project-wide assumptions such as the interest rate for discounting future costs to the present, the average traffic levels at the bridge's location, and the value of driver's time;
- input and edit individual costs for each alternative bridge;
- test to see if the results are sensitive to changes in particular parameters or costs; and
- print reports documenting the steps in the analysis and the results obtained.

To access a step in the left-hand panel, double-click the mouse on the step.

In the center portion of the window is the table of current life-cycle costs by alternative (across the top) and by cost type (down the left). In the upper-right section, the **Edit costs of alternatives** box contains pushbuttons for accessing two project alternatives and creating four additional alternatives. The top row in the table, **Total ($)**, lists the sum of costs currently entered for each alternative. Below this line, the table divides this total three ways, according to three groups: **Costs by bearer**, **Costs by timing**, and **Costs by component**. *The sum of the cost categories under each heading equals the sum listed on the Total ($) line.*

The check boxes in the **Costs by bearer**, **Costs by timing**, and **Costs by component** categories allow the user to display results for a subset of costs. For example, to show only the engineer's estimates of these two structures, the user checkmarks the **Agency** box in the **Costs by bearer** group, the **Initial Construction** box in the **Costs by timing** group, and all four **Element** boxes in the **Costs by component** group. The **Cost Summary** window displays only the engineer's estimates for each alternative bridge, as a total in the **Total ($)** line and by cost types in the three major cost categories. *Note: all subsequent windows, graphs, and reports will display and act on only this subset of costs.*

The upper left box contains **Go Advanced** and **Set as default** buttons. These allow the user to switch back and forth between two fundamental modes in BridgeLCC: the *Basic Mode* that allows the user to conduct and complete analyses without any uncertainty in parameters and the *Advanced Mode* that performs risk and uncertainty analysis. For many if not most analyses, the user can stay in the Basic Mode. (See Section 5.1 for a description of their use.)

The inflation rate, real discount rate, and nominal discount rates are also listed. The inflation rate is used by BridgeLCC to compute the costs of future bridge activities, and the real discount rate is used to compute the present value of these future costs. The nominal discount rate is the combined effect of the inflation and real discount rates. (See the Section 2.3 for a more detailed description of these rates.)

## 1.2    The Economic Foundation of BridgeLCC

BridgeLCC is based on the life-cycle costing, value engineering, and uncertainty analysis techniques described in *ASTM Standards on Building Economics*[1] and on the cost classification scheme developed by Ehlen and Marshall (1996).[2] Competing construction designs, strategies, and materials are assessed by comparing their life-cycle costs when performing the same particular task, say, when constructing, maintaining, and eventually disposing of a two-lane highway bridge. This life-cycle cost (LCC) model shows for each alternative all of the relevant costs for the given function. The alternative that performs the function for the minimum life-cycle cost is the economically efficient choice, other things being equal. The model uses ASTM-standard formulas for discounting future costs to their present-value equivalents and for conducting sensitivity and uncertainty analysis. Additional formulas outside of the *ASTM Standards* are provided for computing the costs to drivers during bridge construction and repair activities (but users can still specify their own user costs).

In BridgeLCC's Advanced Mode, the user can perform Monte Carlo simulations of probabilistic life-cycle cost outcomes based on uncertainty about the value and timing of a project cost, parameters such as the real discount rate, and workzone costs. The user chooses the relative uncertainty of each value; for example, if the best-guess value of a unit cost is $100 per square meter but the cost varies uniformly between $90 and $110, the user can input a unit cost of $100, specify a uniform distribution with an uncertainty of ±10% (e.g., $100 − 0.10($100) = $90). The Monte Carlo simulations produce distributions of life-cycle cost, allowing the user to see the range of life-cycle costs that can result from the uncertainty in costs and parameters. (See Chapter 5 for details and instructions on conducting Monte Carlo simulations.)

---

[1] American Society of Testing and Materials, *ASTM Standards on Building Economics*, Fourth Edition, Philadelphia, PA, 2001.
[2] Ehlen, Mark A., and Marshall, Harold E. 1996, *The Economics of New-Technology Materials: A Case Study of FRP Bridge Decking*, NISTIR 5864, Gaithersburg, MD: National Institute of Standards and Technology.

## 1.3    Organization of this Manual

The manual begins with two chapters that describe the basic functions and options in BridgeLCC. It follows with two chapters describing advanced and optional features and then two chapters describing two example analyses.

Chapter 2 provides a description by function of how to perform a life-cycle cost analysis using BridgeLCC, including how to start an analysis, how to input project parameters and alternatives' costs, and how to produce life-cycle cost tables, graphs, and printed reports. Chapter 3 highlights additional tools that make analyses more comprehensive and productive. Chapter 4 describes how to conduct sensitivity analyses; Chapter 5 covers uncertainty and risk analysis. Chapter 6 describes in detail a basic analysis (most of the figures in Chapters 2 and 3 are taken from this analysis). Chapter 7 follows with more comprehensive analyses that use the advanced uncertainty and risk features described in Chapter 5. Finally, Chapter 8 gives some summary descriptions of additional example analyses included with BridgeLCC. Appendix A outlines the life-cycle methodology in BridgeLCC; Appendix B lists the BridgeLCC discounting and workzone-related user cost formulas.

This manual assumes that the reader has a working knowledge of how to use Microsoft® Windows®, including how to find files and copy them, and to print results to a printer. If you are not familiar with Windows®, please consult a Windows® users guide or operating system manual before proceeding.

The BridgeLCC users manual follows standard window nomenclature, examples of which are shown in Figure 2.

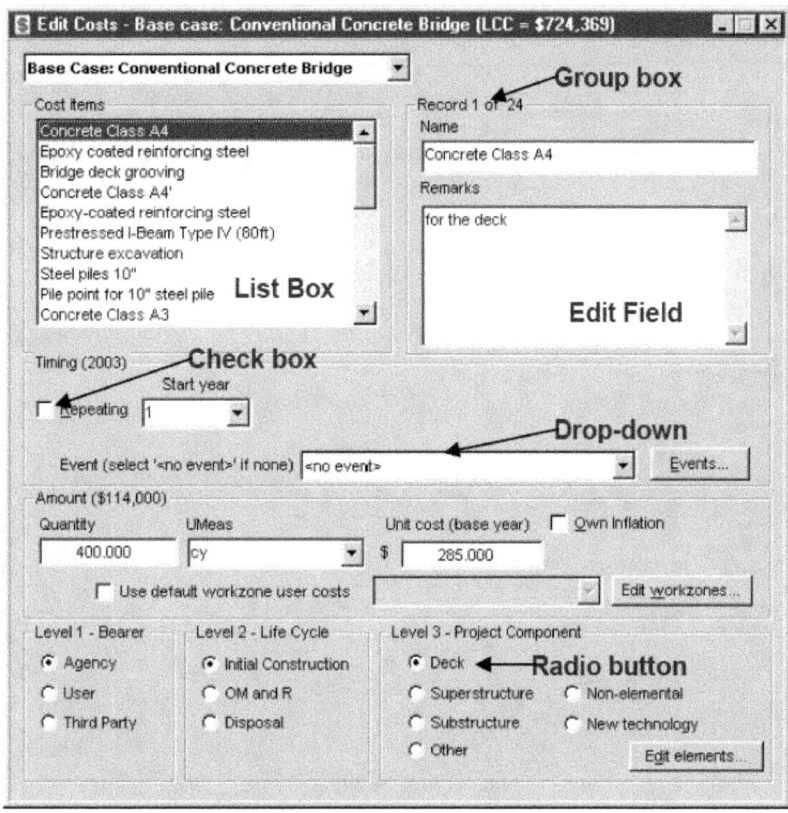

**Figure 2. Standard Window Nomenclature**

Additionally, the manual uses some typographical conventions. Menu and window items are shown in **bold**, and menu items in a task are separated with "/" slashes; for example, to save an analysis, the user is asked to select **File/Save Work...**, that is, to first click **File** on the menu bar and then click **Save Work...** from the **File** submenu. Similarly, if the user is asked to push a window button labeled "OK," the user is asked, "push the **OK** button." The names of the windows themselves are also in bold; e.g., the window in Figure 2 is the **Edit Costs** window. Filenames are shown in quotations; for example, one of the analysis files that comes with BridgeLCC is "Route40.lcc."

| Special notes | Special notes are highlighted with a box similar to the "Special notes" box to the left of this paragraph. These notes provide useful hints and warnings about the proper use of BridgeLCC. |
|---|---|

Help is provided in two forms. First, the software includes online help, which can be accessed by either pressing the **F1** key in the window about which you need help, or by selecting **Help/Topic...** from the menu. Second, this manual is provided with BridgeLCC in the form of an Adobe® Acrobat® PDF file, called "UsersManual.pdf"; it is one of the choices in the BridgeLCC directory in the **Programs** section of your Windows **Start** menu. (For information about Adobe® PDF files and how to install Adobe® Acrobat® on your computer, access the Adobe® web site, http://www.adobe.com.)

# 2. Using BridgeLCC

This chapter describes the three functional tasks necessary to perform a BridgeLCC life-cycle cost analysis: (a) starting an analysis; (b) inputting and editing project data; and (c) computing, interpreting, and reporting the life-cycle cost results. Each function is described with the first-time user in mind and is supported with pictures of BridgeLCC screens and specific instructions on how to accomplish the tasks. Once these functions are understood, read Chapters 3 and 4 for a description of additional BridgeLCC capabilities.

## 2.1    Installing and Starting BridgeLCC

BridgeLCC can be installed by either downloading the software from the BridgeLCC web site, http://www.bfrl.nist.gov/bridgelcc, or installing from CD. The software is designed to run in Windows® 95, 98, 2000, NT, and XP on a computer that has at least a 600MHZ Pentium-level processor, 64MB of RAM, 30MB of available hard disk space, and a video card that supports 1024x768 resolution.

### Installing

When installing from the Internet, simply double-click the downloaded file. This will create a set of files; double-click on the file "setup.bat." When installing from CD, double click the "setup.bat" file on the CD. The installation program will then walk you through a series of screens, including one that allows you to select where on your hard drive BridgeLCC will be installed.

| Note | 1. You need to reboot your computer after you install BridgeLCC 2.0. <br> 2. BridgeLCC is designed to run on a computer with a local hard drive; it will not work if it is installed on any of a computer's remote, networked drives. |
| --- | --- |

To remove BridgeLCC from your computer, access the Windows® **Start** menu and select **Uninstall BridgeLCC** from the BridgeLCC directory. After uninstalling, some files will remain in the BridgeLCC directory on your hard drive, including the "*.lcc" project files you created.

### Starting

Like most Windows programs, BridgeLCC can be started at least two ways. First, it can be started by selecting **Start/Programs/BridgeLCC/BridgeLCC 2.0** from your Windows **Start** menu. Second, it can be started by double-clicking on a BridgeLCC "*.lcc" analysis file located on your desktop or in a folder (for this to work you must have rebooted your computer since installing BridgeLCC 2.0). BridgeLCC will initially greet you with the **Welcome** screen shown in Figure 3.

**Figure 3. Welcome Screen**

## Basic vs. Advanced Analysis

To simplify the process of inputting data, conducting analyses, and interpreting and reporting results, BridgeLCC 2.0 operates in two modes. The first, called the *Basic Mode*, provides the framework and tools for conducting deterministic analysis only. The user inputs his or her best-guess values for parameters, the timing and level of costs, and even the probability of a particular event happening (such as an earthquake). BridgeLCC then computes for each alternative the single-value deterministic life-cycle cost (or in the case of probabilistic events such as earthquakes, its single-value expected life-cycle cost). The second, called the *Advanced Mode*, provides data fields and tools for conducting comprehensive probabilistic analyses.

To see what mode you are currently in, look in the upper left hand corner of the **Cost Summary** window; there will be a box stating either **Current mode: Basic** or **Current mode: Advanced**. To change from one mode to the other, press the corresponding **Go Advanced** or **Go Basic** button; this will replace all open screens with their other-mode counterparts. No data is lost when switching from one mode to the other. The default starting mode can be changed in the **Preferences** window (see Section 3.6 for details).

It is recommended that analyses be carried out in two phases, corresponding to the two modes. First, in the Basic Mode, input your best-guess values for the values and timing of costs, the real discount rate and other parameters, and other values required by your analysis. Compute deterministic life-cycle cost, display graphs showing summary costs and cost timelines, and print reports describing your deterministic analysis of your project. In many cases the Basic Mode, deterministic analysis is sufficient for analyzing your project. If so, then your analysis has been completed.

If, however, you do need to conduct risk and uncertainty analysis on certain values (such as the real discount rate, costs, events, or workzone costs), then switch BridgeLCC to the Advanced Mode (by pressing the **Go Advanced** button), input uncertainty distributions and values in the parameter, cost, events, or workzone fields, and conduct Monte Carlo simulations of the alternatives' life-cycle costs. (See Section 5.1 for more details on using the Advanced Mode.)

**Additional Tips on Navigating through BridgeLCC**

BridgeLCC provides standard tools for navigating through menus and tasks. In addition to the menu bar at the top of the screen:

- use the task bar on the left side of the **Cost Summary** window to access the main windows and associated tasks;

- click the right mouse button on the list boxes in the **Edit Costs** window, **Edit Events** window, and **Workzones** window to add and delete costs, events, and workzones; and

- click the right mouse button on graphs of data to access "pop-up" menus with options for changing colors or graph titles, for printing the graph, and for copying the graph to the clipboard so that it can be pasted in word-processing and presentation programs.

More details about navigating through particular windows can be found in the windows' corresponding sections in this manual.

**2.2    Starting an Analysis**

**Opening an Existing Analysis**

To open an existing analysis, either select **Open existing analysis** from the **Welcome** window (which displays by default when you first start BridgeLCC), or select **File/Open Existing Analysis...** from the menu (when the **Welcome** window has been disabled). Both display a dialog window for accessing and opening "*.lcc" files (Figure 4).

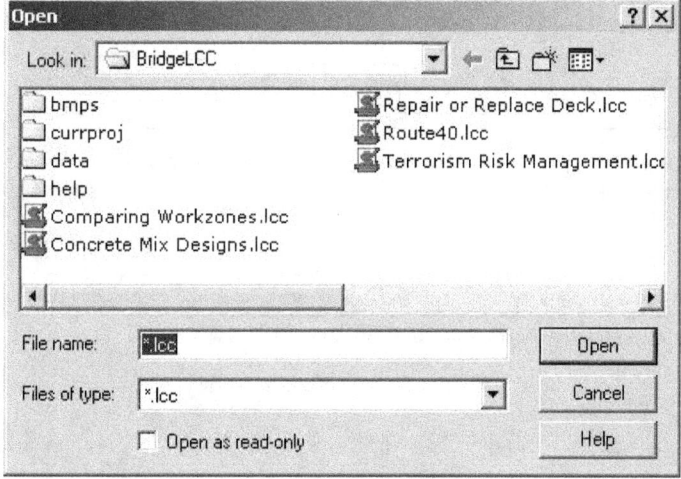

**Figure 4. Open Window**

Once a file has been opened, BridgeLCC will compute the life-cycle costs in the analysis and display them in the **Cost Summary** window.

**Starting a New Analysis**

Start a new analysis by either selecting **Start new analysis** from the **Welcome** window or selecting **File/New Analysis...** from the menu (when the **Welcome** window has been disabled). Only one analysis file can be open in BridgeLCC at a time, so you will need to close an open analysis before starting a new one (by selecting **File/Close Analysis** from the menu).

BridgeLCC will next walk through a series of windows for inputting the minimum information required to start an analysis (Figure 5 through Figure 8).

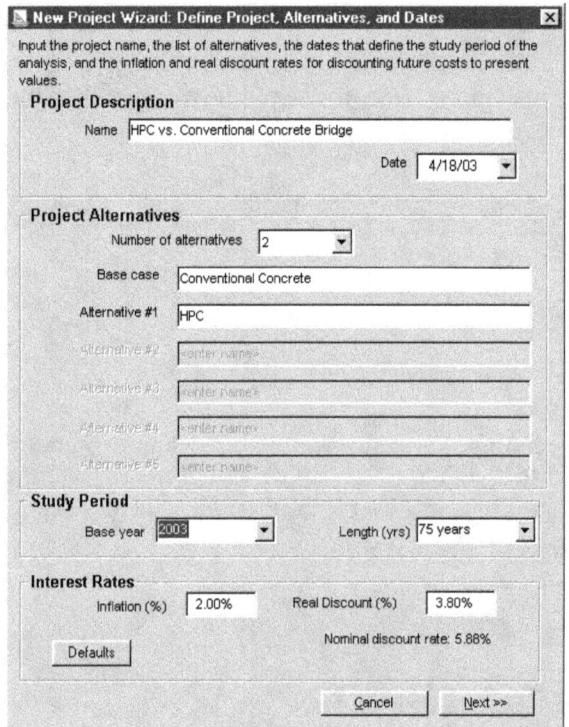

**Figure 5. New Project Wizard – Step 1 of 4: Define Project, Alternatives, and Dates**

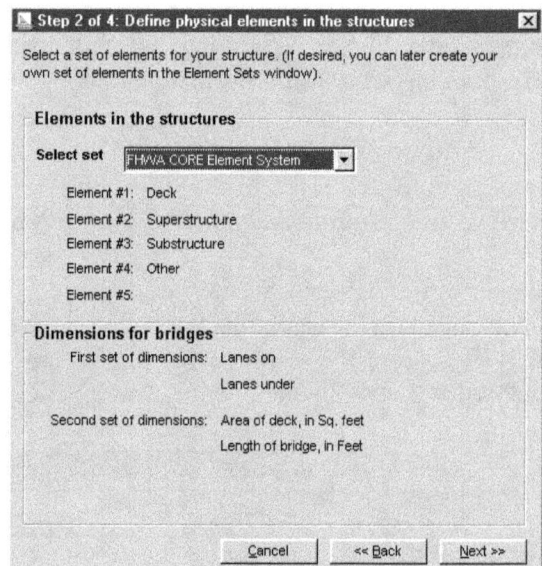

**Figure 6. New Project Wizard - Step 2 of 4: Define Physical Elements**

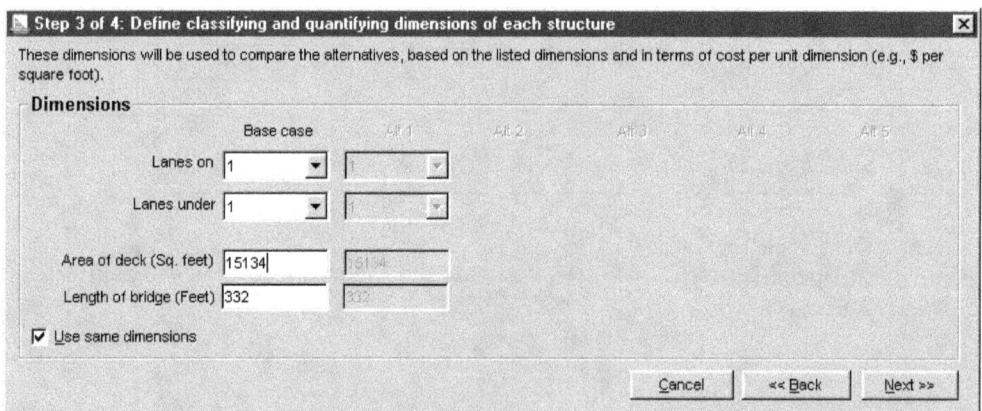

**Figure 7. New Project Wizard - Step 3 of 4: Define Dimensions**

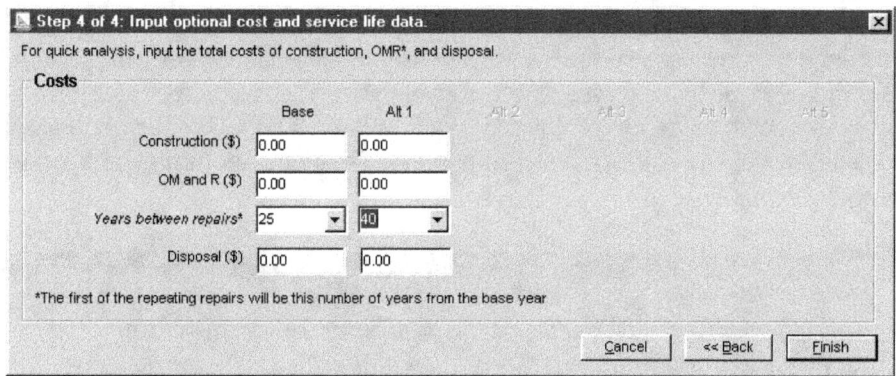

**Figure 8. New Project Wizard - Step 4 of 4: Costs and Years to Repair**

Use the first window to input the name and date of the project, the names of the alternatives, and study period parameters. Use the second window to select the set of project components (e.g., bridge deck) to be used in classifying costs. Use the third to input bridge size information (e.g., length, area) and, at the user's option, the fourth to input simple, summary estimates of construction, maintenance/repair, and disposal costs. Once the four screens have been completed and the **Finish** button in the last has been pressed, the minimum skeleton of an analysis has been created. The **Cost Summary** window will then display with the inputted data and current life-cycle cost totals for each alternative.

**Cost Summary Window**

The **Cost Summary** window serves three important functions:

1. it summarizes the current LCC calculations, by listing totals, totals per unit (e.g., square meter of bridge deck), and net savings;

2. it can filter costs and alternatives so that all calculations, graphs, and reports act on only the filtered subset; and

3. it provides a list of the steps required to complete a life-cycle cost analysis (that double as a means for accessing the windows that carry out these steps).

**Steps in a Life-Cycle Cost Analysis**

BridgeLCC uses the ASTM E 917 practice, in which, for example, the user defines the project, the alternatives, and the costs, and carries out a sensitivity analysis. The menu panel on the left of the **Cost Summary** window reflects these ASTM E 917 activities. By double-clicking the mouse on an item on the list, the user can access the window(s) needed to complete the task. For example, the user can double-click **Description** to access the **Project Description** window and edit the project objective.

11

## Current State of Life-Cycle Cost Calculations

The **Cost Summary** window also summarizes the current state of the cost calculations. The middle portion of the window lists, by cost type, the life-cycle cost of each project alternative. In Figure 9, these alternatives are listed as **"Conve"** for the base case and **"HPC B"** for alternative #1 (the two words are abbreviations of the alternatives' full names). The numbers in parentheses are the number of costs currently in each alternative.

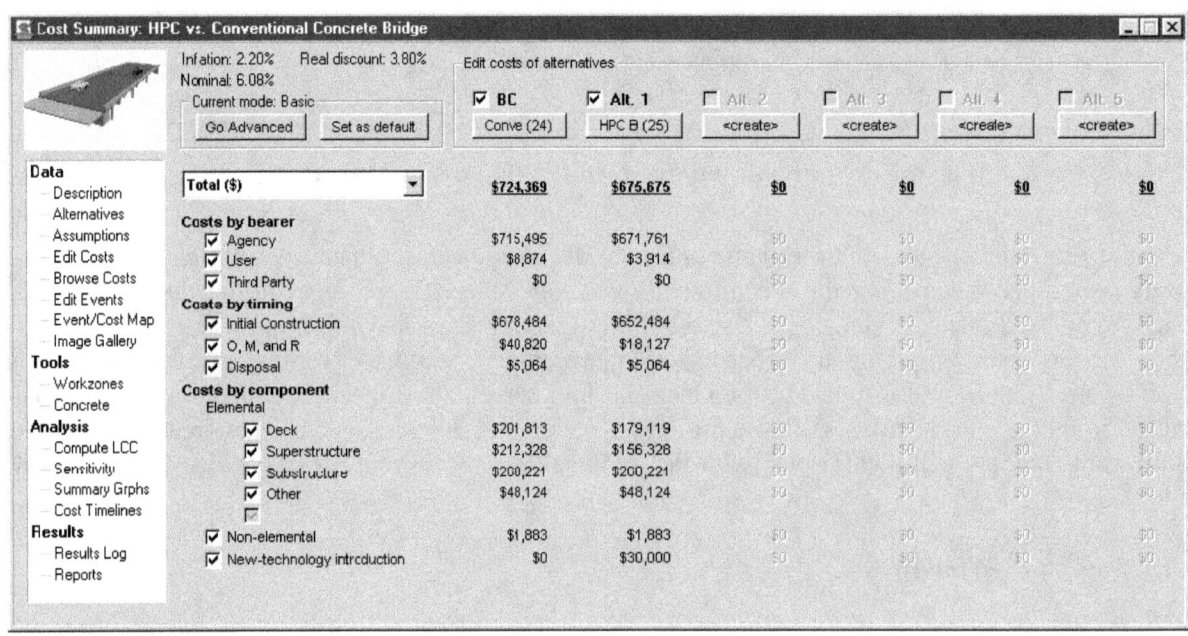

Figure 9. Cost Summary Window

The **Total ($)** line toward the top lists the total life-cycle costs for each alternative, while the **Costs by bearer**, **Costs by timing**, and **Costs by component** groups show three different breakdowns of this total. For example, the current total for Alternative #1, "HPC," is $675,675. Looking at the **Costs by bearer** categories, $671,761 of this cost is Agency costs (costs incurred by the agency that builds, maintains, and disposes of the structure), and the remaining $3,914 is User costs (costs incurred by drivers on the bridge).

> **Note** The sum of cost categories within each of the three **Costs by** categories <u>always</u> equals the **Total ($)** value displayed. BridgeLCC recalculates life-cycle costs after each edit to a parameter or value that affects these costs.

Similarly, looking at the **Costs by timing** breakdown, of the $675,675 total, $652,484 occurs during initial construction; $18,127 occurs during operation, maintenance, and repair (O, M, and R); and $5,064 occurs during disposal of the structure. Finally, looking at the **Costs by component** breakdown, of the $675,675 total, $179,119 is associated with the deck, $156,328 is associated with the bridge superstructure, $260,221 is associated with the bridge substructure, $48,124 is associated with other physical components with the structure, $1,883 is associated with non-elemental parts of the bridge (e.g. overhead costs, mobilization), and $30,000 is associated with "new-technology" activities (i.e., introducing and using the material for the first time).

To see the life-cycle costs expressed in dollars per unit of bridge (e.g., per square meter or lineal meter of bridge deck), select the **Total ($)** drop-down box and select **$ per <bridge unit>**. In our example, our bridge unit choices are **$ per square feet** and **$ per feet.** To see the results expressed as net savings over the base case, select **Net Savings ($ per <bridge unit>)**. Positive values for an alternative indicate net savings over the base case; if the values are negative, then the alternative's costs are greater than the base case.

### Filtering by Cost Type and Alternative

Since life-cycle costs involve summing costs that are incurred by different groups, over different periods of time, and in different parts of the project, it is useful to sometimes look at, edit, and perform an analysis on only a subset of all costs and alternatives. For example, the user may want to compare only the **Agency, Initial Construction**, and **Deck** costs of the **Base Case** and **Alternative #1**. By check-marking only the **Agency, Initial Construction**, and **Deck** checkboxes and only the **Base Case** and **Alternative #1** alternatives, BridgeLCC will display only these cost types and alternatives and generate graphs and reports that display data on these cost types and alternatives.

| Technical Note | The life-cycle cost of an alternative is the sum of <u>all</u> costs associated with the alternative, not of any <u>subset</u> of these costs. So while it is useful to filter costs for the sake of inputting and editing values, and for explaining results, the <u>life-cycle cost-effective alternative</u> is the one with the lowest sum of <u>all</u> costs. |
| --- | --- |

To checkmark all of the **Costs by bearer, Costs by timing**, and **Costs by component** categories of cost, position the mouse over one of the check boxes and click the right mouse button; a "pop-up" menu will appear with options to checkmark all of the buttons (as well as to uncheck them all).

## 2.3     Inputting Project Data

After creating an analysis, the next functional step is to input analysis data. Using the ASTM standard practice as a guide, there are three basic types of data:

1. **project description and alternative data** – at least the minimum performance criteria of the project and the alternatives under consideration that meet these criteria;

2. **project assumptions** – the parameters common to all alternatives, including the discount rate and the traffic parameters used to calculate user costs; and

3. **cost data** – the individual costs that make up each project alternative's life-cycle cost.

BridgeLCC also allows the user to create and edit additional types of data, such as "events"; these optional features are discussed in Chapter 3.

### Project Description and Alternatives

Data for the project and its alternatives are entered into the **Project Description** and **Alternatives** windows. The **Project Description** tab (Figure 10) is used to name the project, to date it, and to describe its objective and the performance requirements that each alternative must meet. Be sure to give an accurate description of the performance requirements each alternative must satisfy (e.g., HS-20 loads and a 75-year life). Press the **Gallery** button in the bottom right corner to access the **Image**

**Gallery** and to view/update the list of images showing characteristics of this project (such as photos of current conditions).

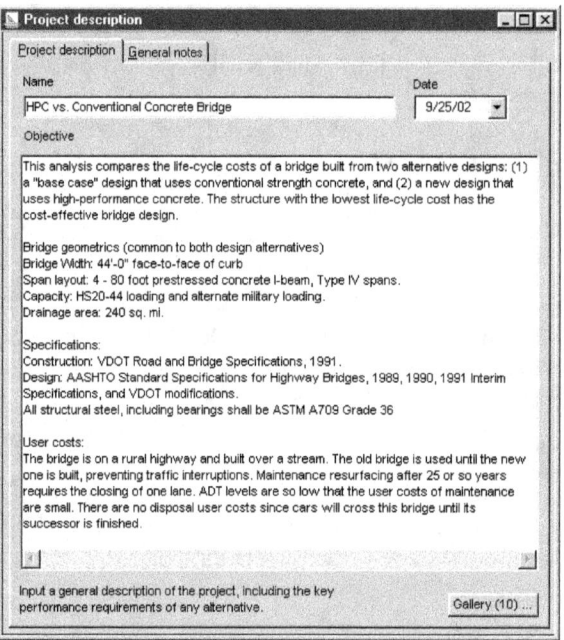

**Figure 10. Project Description Tab**

The **Project Description** window can also be used to document notes about the basis of this analysis (why and for whom it is being performed) and to summarize the results (Figure 11). Both are highly visible parts of the BridgeLCC reports that are designed to be submitted to stakeholders and clients.

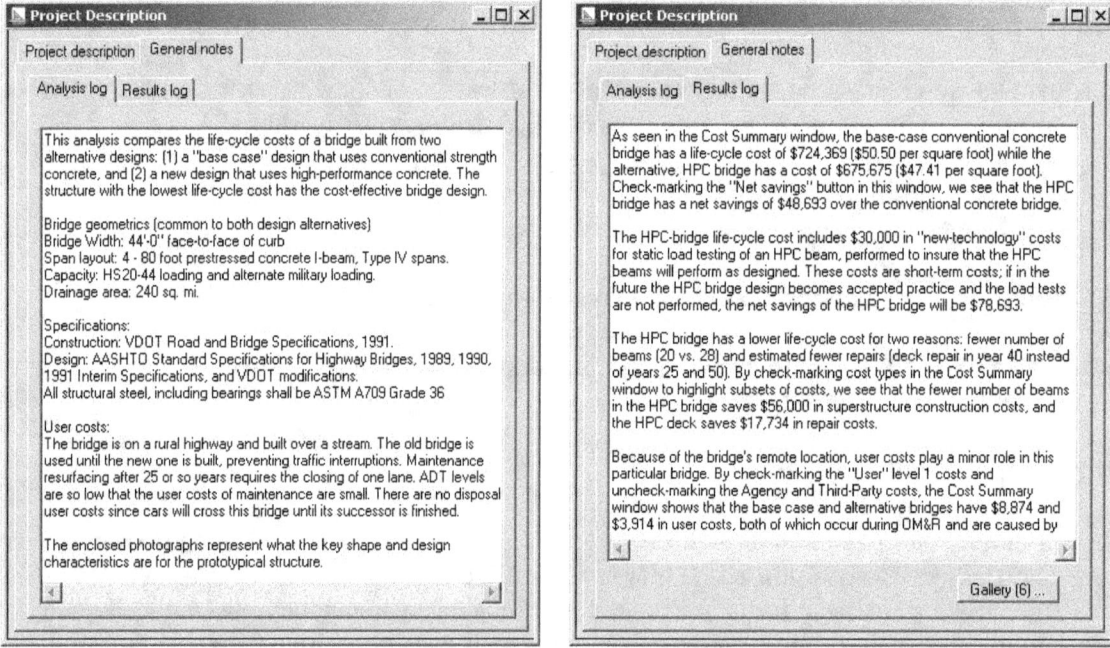

**Figure 11. General Notes Tabs**

Project alternatives are created and edited in the **Alternatives** window (Figure 12).

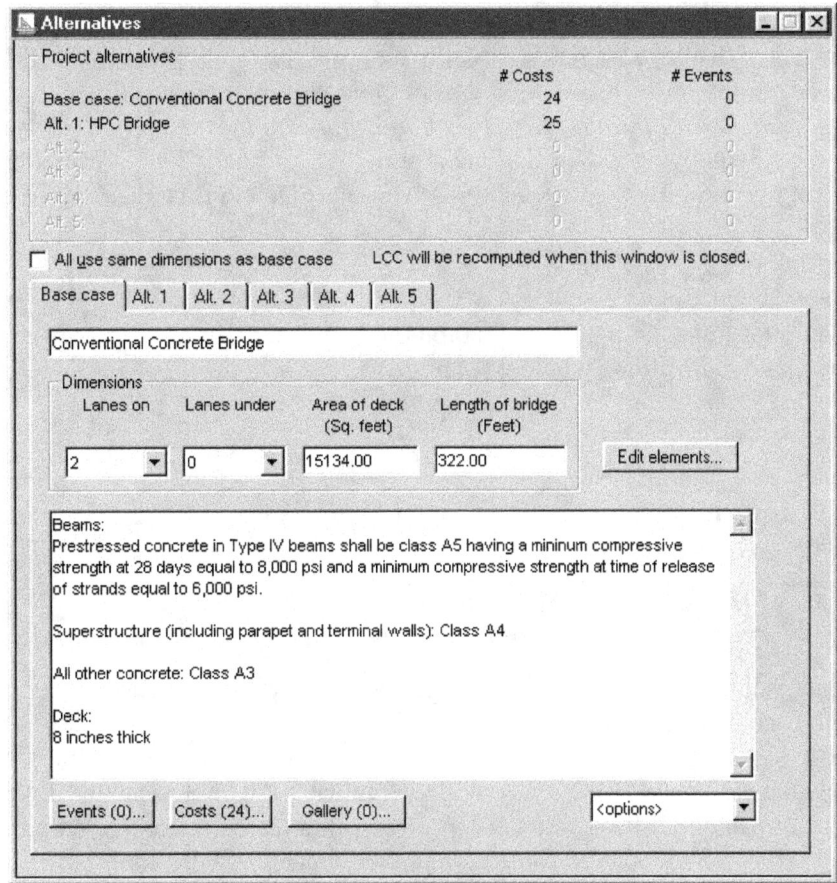

**Figure 12. Alternatives Window**

The top of the window lists the current set of alternatives, including the total number of costs and events in each. To create an alternative, access the particular alternative's tab and press its **Create alt** pushbutton. Input values for the classifying dimensions (in Figure 12 they are **Lanes on** and **Lanes under**) and the quantifying dimensions (in Figure 12 they are **Area of deck (Sq. feet)** and **Length of bridge (Feet)**). Classifying dimensions are used for comparing the results of different LCC analyses according to bridge characteristics – bridge lanes in this case, which quantifying dimensions are used to compute life-cycle costs per unit, such as life-cycle costs per square foot of bridge deck. If you want all of the alternatives to have the same classifying and quantifying dimensions, then input your dimensions in the **Base case** tab and then checkmark the **All use same dimensions as base case** check box. Classifying and quantifying dimensions are set in the **Elements** tab of the **Project Assumptions** window (Element sets and classifying and quantifying dimensions are covered later in this section).

The large text area is used to document the characteristics of the particular alternative (design, construction techniques, maintenance requirements), particularly those characteristics that are different from the other alternatives and that will impact life-cycle costs. The **Events()...**, **Costs()...**, and **Gallery()...** buttons toward the bottom provide access to the event, cost, and image windows for

15

this alternative; they also display in parentheses the current number of each type, for example, the label "Costs (24)" indicates that the alternative currently has 24 costs. (The inputting and editing of events is covered in Section 3.1; costs and images are covered in this section under "Project Costs.")

To copy the currently displayed alternative to a new alternative, select **Copy this alt** from the **<options>** drop-down box in the lower-right corner. You will then be asked which alternative is the target. *Note: this will destroy all data in the target alternative.* For example, if you want to copy the Base case to a new Alternative #4, then first make sure that Alternative #4 does not have any needed data in it. To delete the currently displayed alternative, select **Delete this alt** from the **<options>** list.

### Project Assumptions

The project assumptions are the parameters common to all alternatives; they are divided into four groups in the **Project Assumptions** window (Figure 13):

- Economic
- Workzones
- Concrete mix designs (for concrete service life prediction)
- Elements

### Economic Tab

The economic data includes the base year and length of study period, the currency to be used, and the inflation and real discount rates (Figure 13).

The **Base year** is the first year of the study period (typically the first year of construction). It also serves as the year on which all life-cycle cost (present value) calculations are based. The **Length** is the duration of the study period, which is the period over which costs are analyzed.

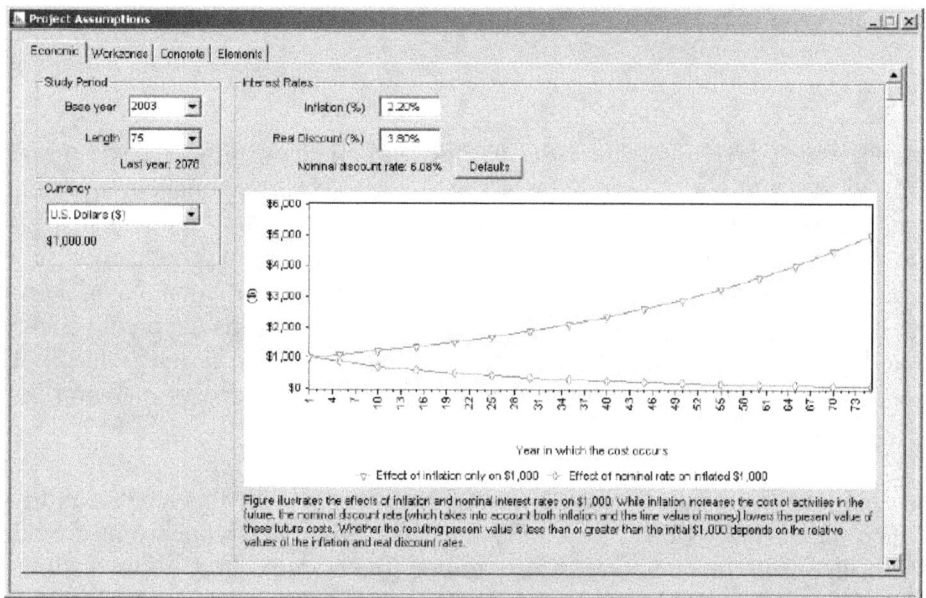

**Figure 13. Economic Data Tab**

The graph of costs over time illustrates the effects of the selected inflation and real discount rates on individual costs. The upper line shows the effect of inflation on costs, in particular, it shows how an activity that costs $1,000 in the base year will, at an inflation rate of 2.20%, increase to over $5,000 in 75 years. The lower line shows the effect of the real discount rate on that inflated $1,000 cost. While inflation will cause this $1,000 cost to be over $5,000 in the 76th year, its present value will be $61. The inflation rate is often set to the Consumer Price Index, produced by the Bureau of Economic Analysis. The real discount rate for Federal infrastructure projects is set by the Office of Management and Budget.

| Current Rates | Current inflation and real discount rates can be found on the BridgeLCC web site, http://www.bfrl.nist.gov/bridgelcc, as well as at their original sources at BEA and OMB. See also NIST's *Energy Price Indices and Discount Factors for Life-Cycle Cost Analysis.*[3] |

The effects of inflation and real discount rates are important for different reasons. The inflation rate determines the sums of money – explicit, and in the case of workzone user costs, implicit – that will be spent over time. Take a simple example: the agencies building and maintaining a bridge need to know how much money to budget for a particular structure. Applying the inflation rate to all costs and viewing the BridgeLCC annual and cumulative current year graphs of agency costs will tell them the physical dollars that will be spent on the bridge over the study period (see Section 2.4 for descriptions of these graphs).

---

[3] Rushing, Amy S., and Fuller, Sieglinde K. *Energy Price Indices and Discount Factors for Life-Cycle Cost Analysis.* NISTIR 85-3273-18. Gaithersburg, MD: National Institute of Standards and Technology, 2003.

The real discount rate is used to compute the present value of all costs, the sum of which is the life-cycle cost of the structure. The real discount rate represents the time value of money; when combined mathematically with the inflation rate, it forms the *nominal discount rate* (often just termed the "discount rate"), which represents both the time value of money and the change in prices of construction and services over time.[4]

## Elements Tab

So that individual costs and life-cycle costs can be classified by project component (e.g., deck, superstructure, substructure), the user defines up to 5 project components. In addition to these physical components, BridgeLCC also supplies a Non-elemental component (for costs not attributable to physical components, such as mobilization and overhead costs) and a New-technology component (for costs incurred due to the first-time use of a new construction material or process, and once the new technology is accepted, not incurred). Engineering firms supplying life-cycle costing reports to state agencies with different elemental breakdowns (some may use the FHWA/PONTIS CORE elements, others may use their own state's elements) may need to maintain several sets of components.

The **Elements** tab allows the user to view and set the element set for this analysis. In Figure 14 the user has selected the "FHWA Core Element System." Its classifying dimensions are the number of **Lanes on** (the bridge) and the number of **Lanes under** (the bridge); these allow the user to compare the results of different BridgeLCC analyses based on the general class of structure (eight-lane bridges should not be compared with single-lane bridges). Its quantifying dimensions are **Area of deck (Sq. feet)** and **Length of bridge (Feet)**. The quantifying dimensions are used in the **Cost Summary** window for listing life-cycle costs per unit (e.g., $ per square foot of deck) and in the **Edit Costs** window for inputting costs.

---

[4] For more details on the use of the inflation rate and real and nominal discount rates, see ASTM, *Building Economics*.

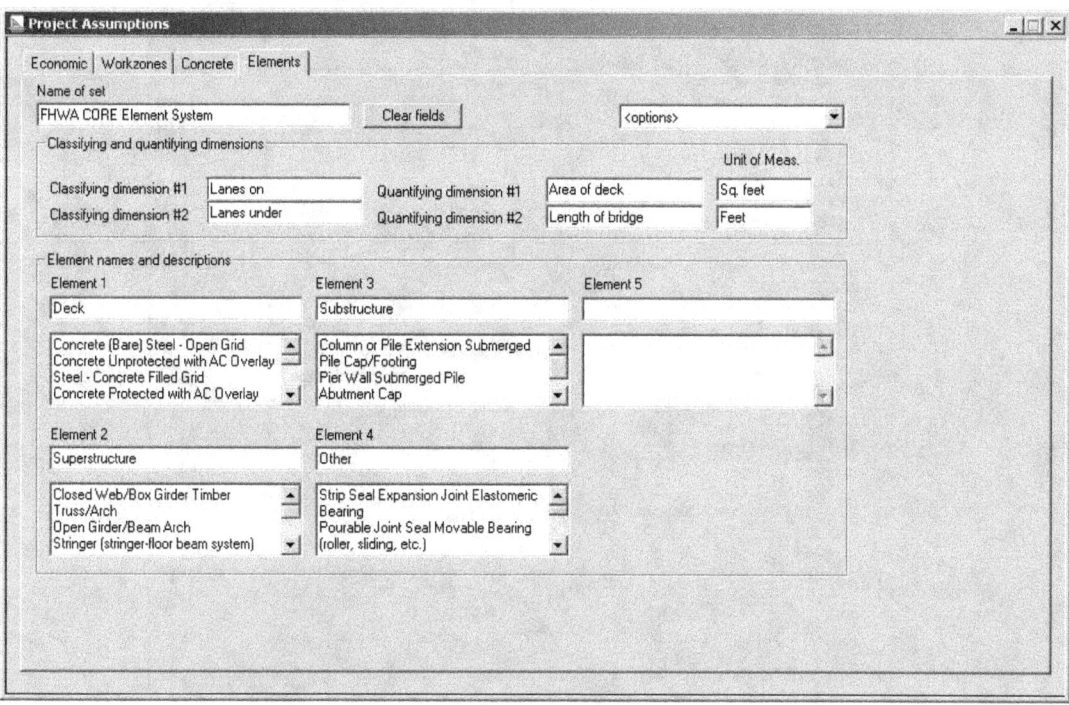

**Figure 14. Elements Tab**

To save the currently displayed element set to the BridgeLCC element database on your machine, select **Add set to BridgeLCC database** from the **<options>** drop-down. You will be presented with a database screen. To import a set from this screen into the **Elements** tab, select **Import set from BridgeLCC database** from the **<options>** drop-down.

The **Project Assumptions** window contains two additional panels, the **Workzones** tab and the **Concrete** tab. Since neither is essential to completing a Basic-Mode life-cycle cost analysis, they are covered in Chapter 3.

**Project Costs**

BridgeLCC provides two windows for inputting and editing alternatives' costs: the **Edit Costs** and **Browse All Costs** windows. Figure 15 shows the **Edit Costs** window.

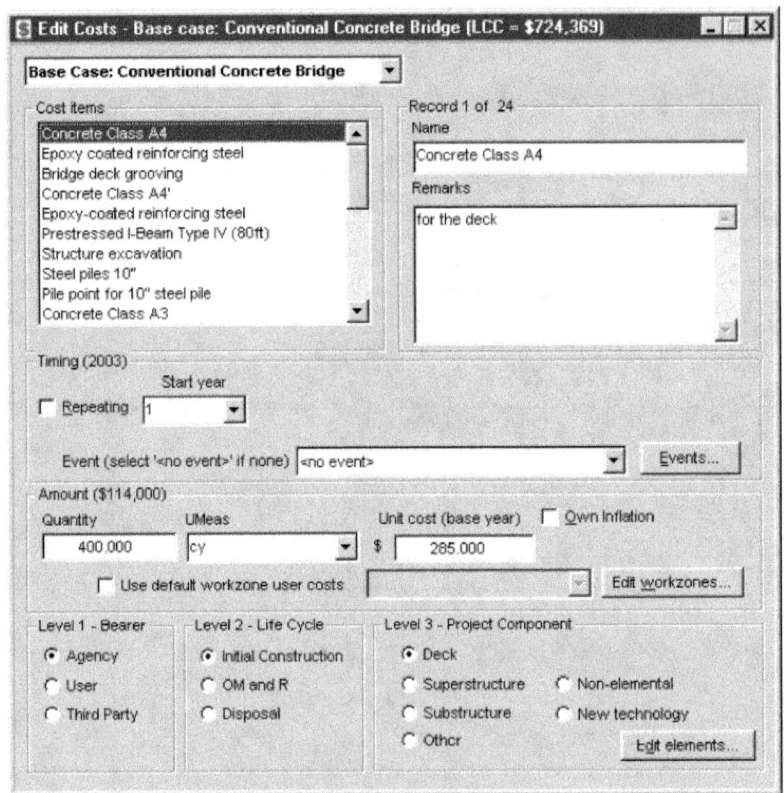

**Figure 15. Edit Costs Window**

Use this window to create, edit, classify, and delete costs. The data on each cost are divided into naming, timing, amount, and classification sections. The **Name** edit field lists the name that will appear on all reports listing individual costs; the **Remarks** edit field is used to list important supporting data such as the source of the cost (e.g., a previous analysis) or anything that limits use of the cost (e.g., cost does not apply to very large structures).

| Note | To add, copy, or delete costs, position the mouse over the **Cost items** list and click the right mouse button; a pop-up menu will appear with these options. |
| --- | --- |

The **Timing** group is used to date the cost. If the cost occurs only once, uncheck the **Repeating** box and select the year in the **Start year** drop-down list. If the cost occurs more than once – for example, every other year for ten years – then checkmark the **Repeating** box, select values in the **Start year** and **End year**, and input a frequency in the **Every ____ years** field, either by selecting a value from the list or inputting a value.

Following the convention of most engineers' estimates, cost amounts are input in terms of quantity, unit of measure, and unit cost. The exception is when inputting a user cost (by selecting the **User** button in the **Level 1 - Bearer** group); in this case the **Amount** group allows you to either (1) use the default values of user costs, based on per-day driver delay, vehicle operating, and accident costs calculated using the traffic parameters specified in the **Project Assumptions** window, or (2) input

20

your own user costs. To specify your own user costs, un-checkmark the **Use default workzone user costs** box.

| Technical Note | Input costs in the **Unit cost (base year)** field in *base-year dollars*, that is, what it costs to perform the task in the base year of the study period. For example, if the base year is 2003 and the agency pays $5 per square foot in 2003 to repair a deck, then $5 is the cost in base-year dollars.<br><br>If in the Year 2004 it costs $6 per square foot for the same repair, then $6 is the cost in *current-year* dollars. The $6 cost would need to be converted to its value in the base year ($5) before being inputted in the **Unit cost (base year)** field. |
| --- | --- |

Use the **Level 1**, **Level 2**, and **Level 3** classification groups to classify the cost according to (1) who incurs the cost (**Agency**, **User**, or **Third Party**), (2) in what period of the life cycle the cost occurs (**Initial Construction, OM&R,** or **Disposal**), and (3) in what part of the project the cost occurs (e.g., **Deck**). All three levels are used in the **Cost Summary** window to list life-cycle costs according to these levels.

Scroll through the alternative's costs by selecting from the list box in the upper left-hand corner. To create a new cost, copy the current one, copy all costs, delete the currently selected cost, or delete all shown costs, position the mouse cursor over the list box in the upper left corner and click the right mouse button.

**Filtering Costs in All Windows**

The user can view a subset of the alternative's costs by going to the **Cost Summary** window and check-marking the desired subset of **Level 1**, **Level 2**, and **Level 3** cost types. For example, to see only the engineer's estimate, checkmark only the **Agency**, **Initial Construction**, and the four **Element** check boxes in the **Cost Summary** window and then return to the alternative's **Edit Costs** window; the list of costs will be updated to reflect this filtering.

**Browsing All Costs in an Alternative**

The second window for viewing and editing alternatives' costs, in a "spreadsheet" form, is the **Browse All Costs** window, shown in Figure 16.

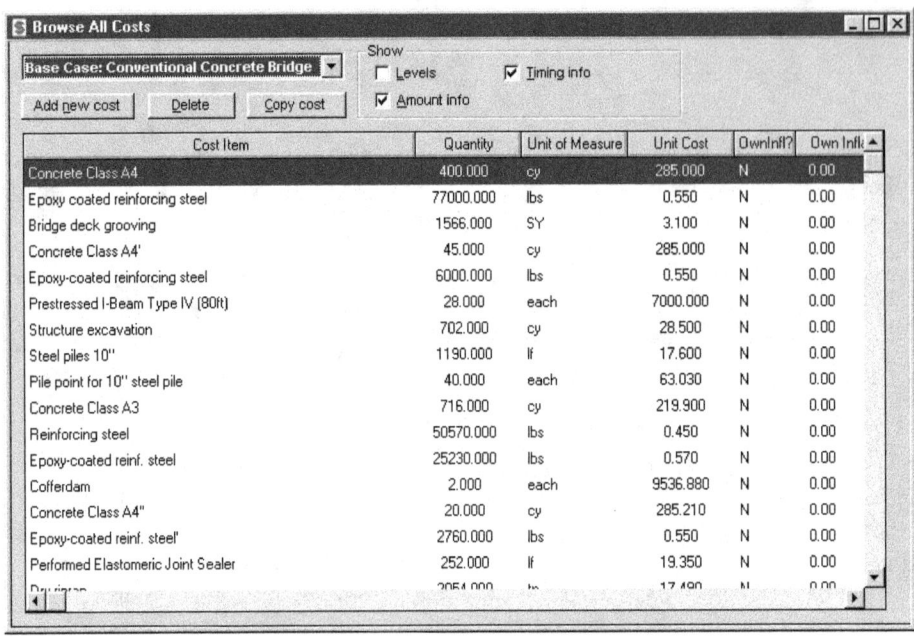

**Figure 16. Browse All Costs Window**

To view a particular alternative, click the alternative in the upper left-hand drop-down box. To change the types of information shown for each cost, checkmark the appropriate boxes in the **Show** group. Edit cost data by double-clicking in the spreadsheet view. New costs can be added and the current cost deleted or copied by pressing the associated buttons below the alternative list box.

### Project Reports

Once the project description, alternatives, parameters, and costs have been inputted, BridgeLCC can print reports so that the user can verify that these data are correct. Select **File/Print...** from the menu to access the **Reports** window. Figure 17 shows this window and the boxes that should be checked to print reports of the data covered in the previous sections.

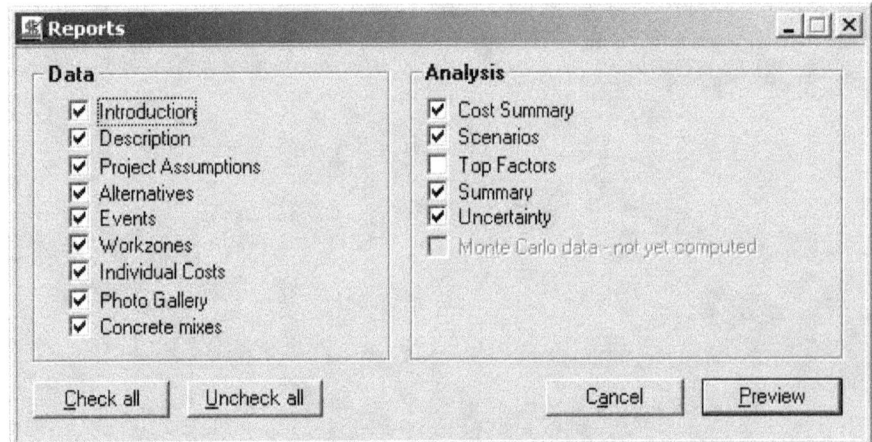

**Figure 17. Reports Window**

## 2.4    Reviewing Results

### Computing Life-Cycle Costs

BridgeLCC computes life-cycle costs in "real time" – with each change in costs, real discount rate, or other parameter, the software automatically re-computes life-cycle costs. So there really is no effort to "computing life-cycle costs," other than verifying the correctness of your data.

### Cost Summary Window

The **Cost Summary** window (Figure 18) is the primary means for reviewing the life-cycle costs of each alternative. The window has some important features for displaying alternative measures of life-cycle cost and for viewing subsets of costs and alternatives.

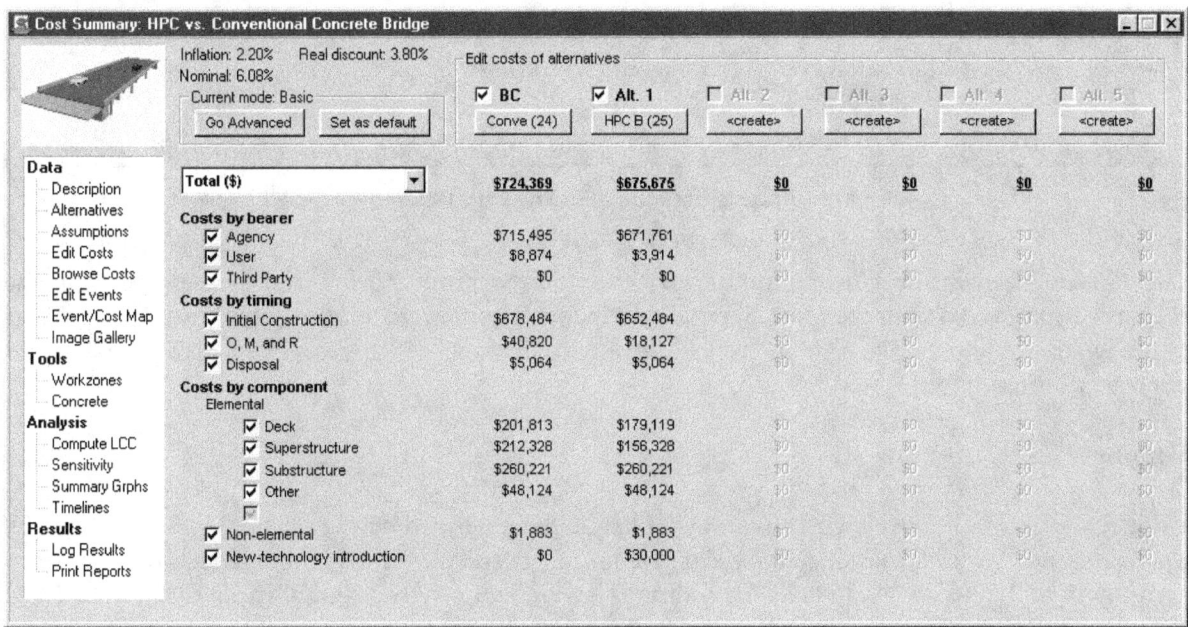

**Figure 18. Cost Summary Window**

In addition to displaying the levels of life-cycle cost, it can display

- life-cycle costs per bridge unit (e.g., life-cycle costs per square meter of bridge deck),
- net savings of the alternative over the base case, and
- net savings per bridge unit.

These alternate measures of life-cycle cost effectiveness can be displayed in the window by accessing the **Total ($)** list box and selecting the alternate measure.

The window can also display results for a subset of costs and of alternatives. For example, if you wish to see only the agency's initial construction costs for the elemental and non-elemental components, checkmark the boxes as they are in Figure 19.

23

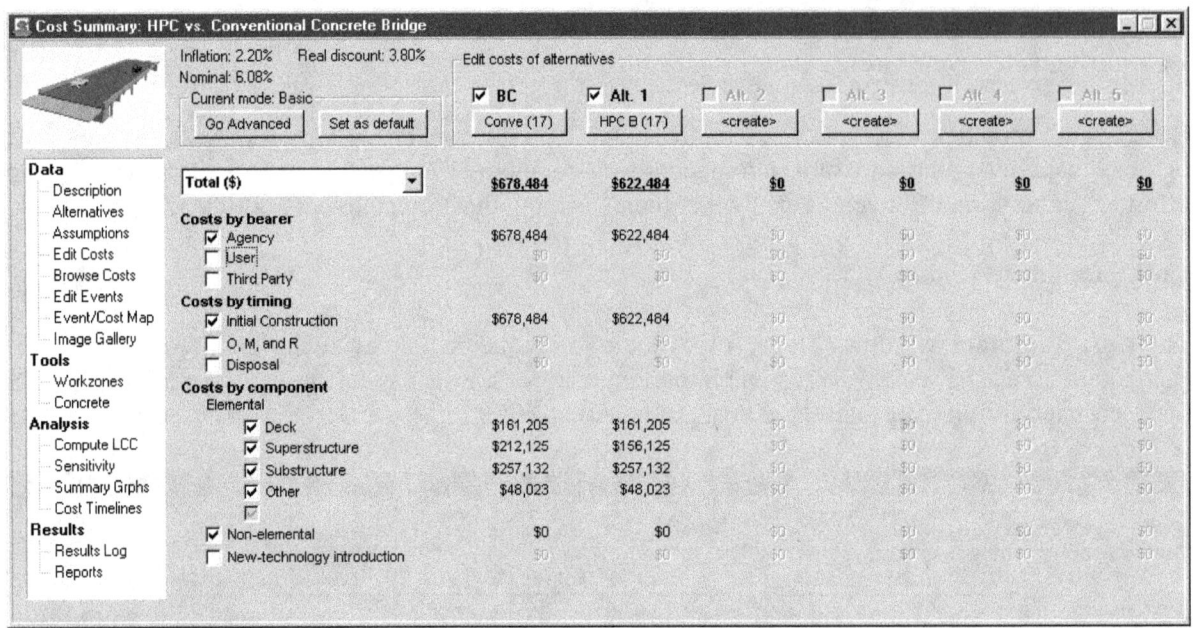

**Figure 19. Cost Summary Window with Filtered Costs**

The figures listed in the **Cost Summary** window can be printed using one of the reports in the **Reports** window. To print the **Cost Summary** window, position the mouse cursor over a blank part of the window and press the right mouse button; a pop-up menu will appear listing an option to **Print current window**.

## Graphs

BridgeLCC graphs life-cycle costs two ways. First, it generates summary graphs of life-cycle cost, by alternative and by cost classification. The graphs can be displayed by either selecting **Summary Graphs** from the **Tasks** list in the **Cost Summary** window or by selecting **Graphs/All Three LCC graphs...** from the menu.

Figure 20 shows the three LCC summary graphs.

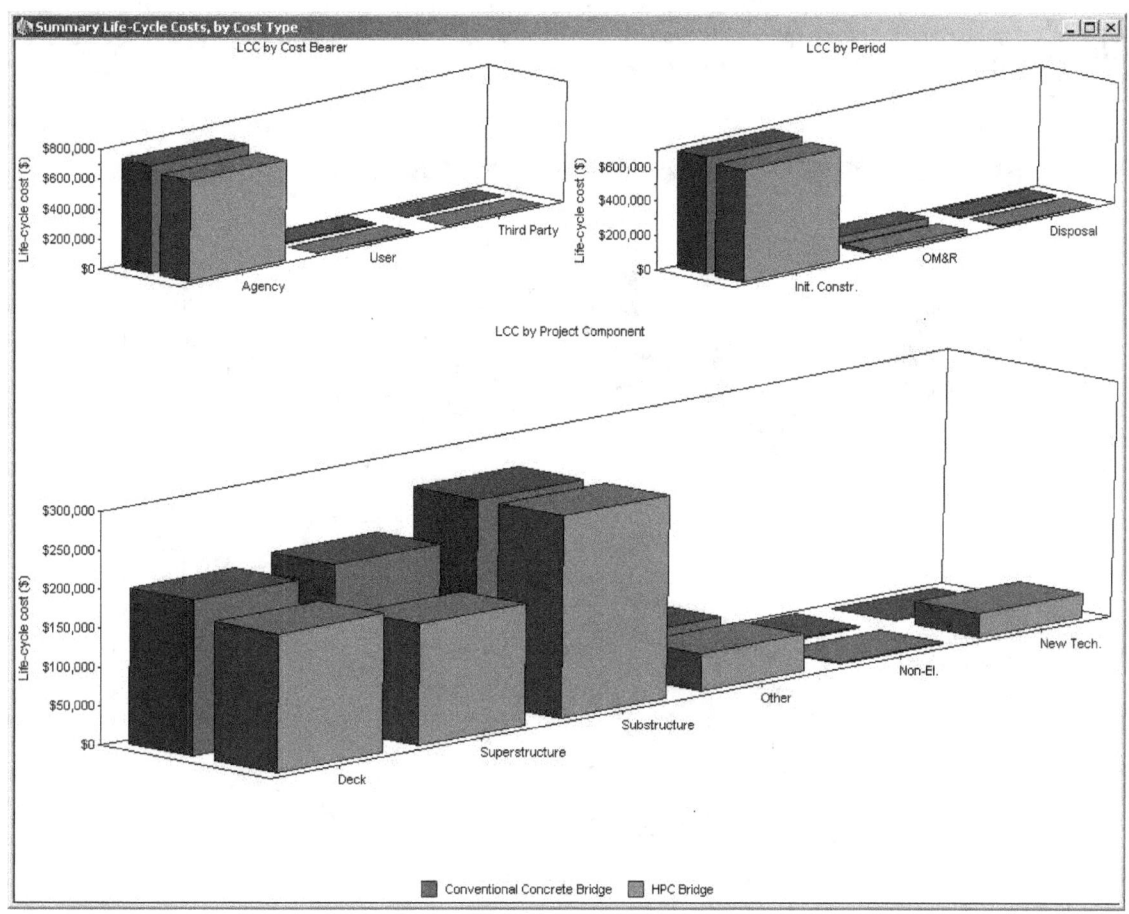

**Figure 20. LCC Summary Graphs**

The second set of graphs that BridgeLCC can generate is timelines of costs. The four different graphs are shown in Figure 21. The top-left graph shows annual costs in current-year dollars (what the costs will be in the particular year) and the graph below it shows the cumulative total of these annual costs. The top-right graph shows annual costs in constant dollars (or present-value dollars) and the graph below it shows the cumulative total of these annual costs.

25

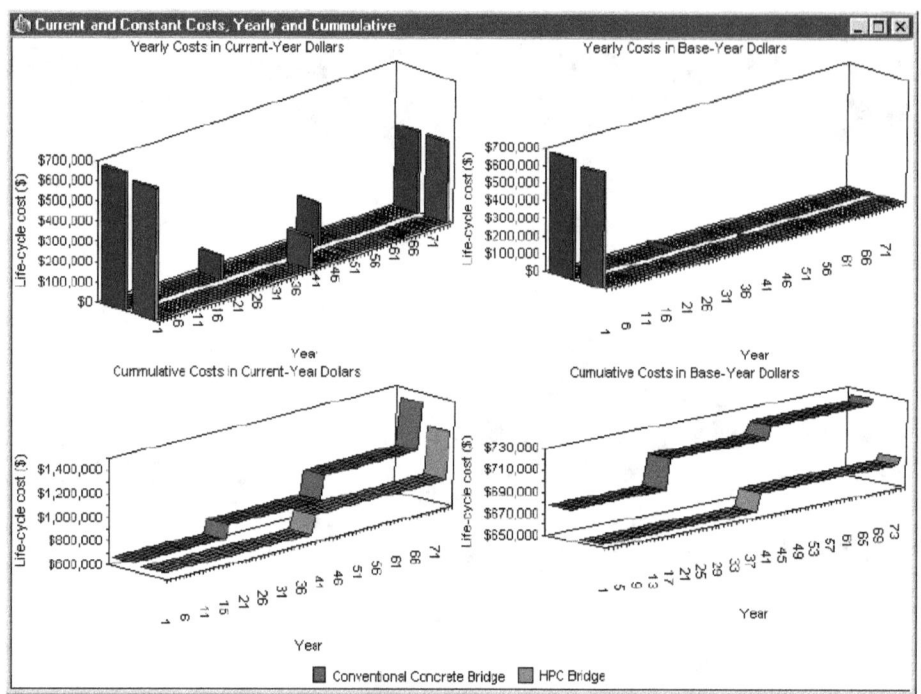

**Figure 21. The Four Timeline Graphs**

Both the LCC summary graphs and the timelines of current-year and base-year costs can be printed, modified, added to the gallery, and copied to the clipboard for pasting in other applications by positioning the mouse cursor over a blank part of a graph and clicking the right mouse button. A menu will appear, listing these options.

## Reports

BridgeLCC prints reports for most of the data and analyses it performs. As shown in Figure 22, the reports are grouped according to whether they display input data or results from the analysis.

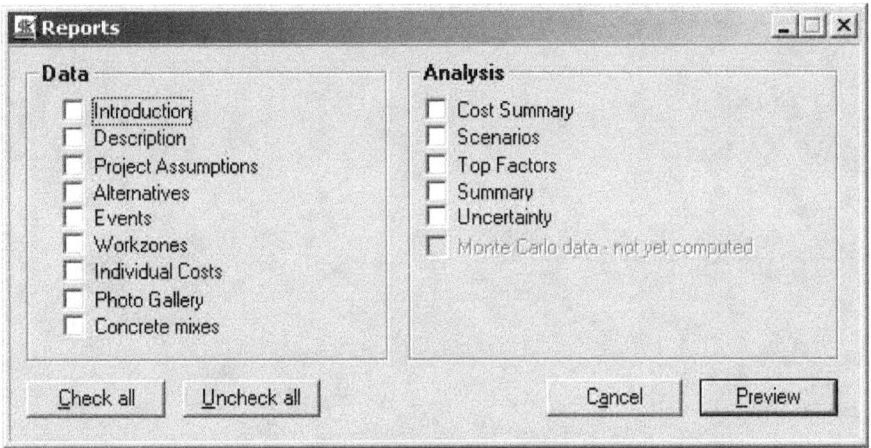

**Figure 22. Reports Window**

After selecting the sub-reports to be printed, press the **Preview** button to see a preview of what the reports will look like (Figure 23 shows an example output). If they look okay, then they can be printed from the preview window.

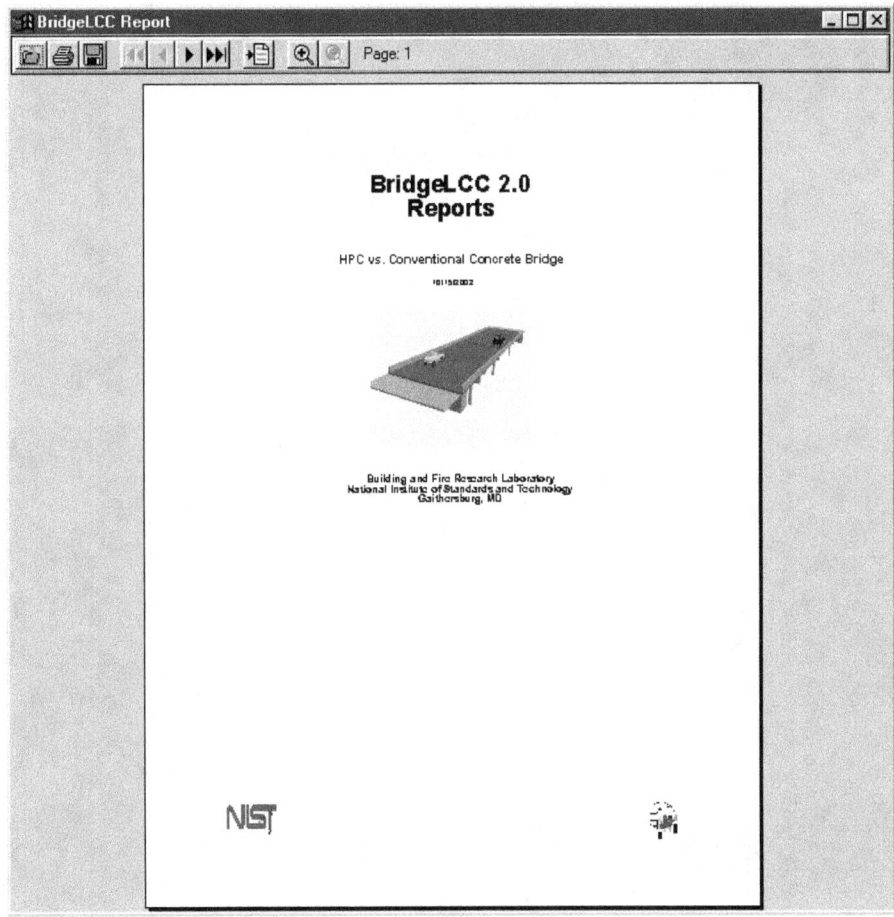

**Figure 23. Sample Report**

## 3. Additional Tools

Chapter 2 described the basic set of tools necessary to conduct a life-cycle cost analysis. This chapter introduces additional tools that can significantly simplify more complicated and comprehensive types of analyses.

### 3.1    Events

Consider a given bridge alternative that requires repairs to its deck every 15 years. Each repair involves numerous steps and costs, each of which must be individually inputted into BridgeLCC. During sensitivity analysis the user would like to know the effect of changing this 15-year repair cycle on life-cycle costs.

The timing of all of the costs associated with the repair can be assigned to a BridgeLCC repair *event*. The sole purpose of BridgeLCC events is to provide a calendar of scheduled events to which costs can be collectively assigned. This event can be recurring or non-recurring, can be specifically defined to occur after another event (e.g., a repair event can occur after a construction event), and can be assigned a probability of occurring. (The **Edit Events** windows in this chapter are from the included example, "Terrorism Risk Management.lcc.")

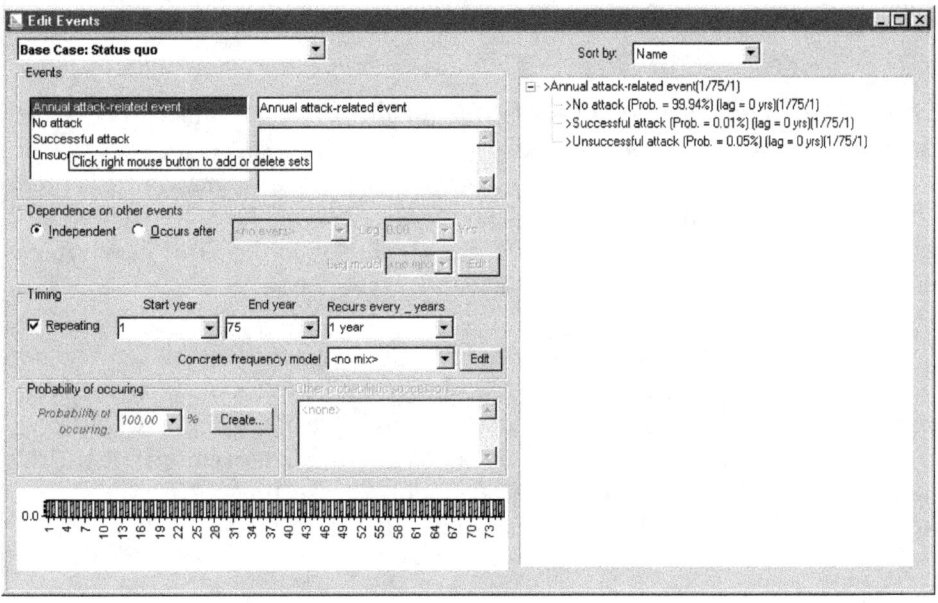

**Figure 24. Edit Events**

**Events**

The left panel lists the complete set of events for this alternative (which is selected in the upper-left list box), and the edit fields for the name and description of the current event. To add, copy, or delete events, position the mouse over the list of events and click the right mouse button; a pop-up menu will appear listing these options.

29

**Event data**

If the event is independent of other events, that is, it does not explicitly occur after another event, select **Independent**. If, on the other hand, the event (such as **No attack**) occurs after another event, there are several settings to make. First, select the **Occurs after** button and select a predecessor event (in this case, **Annual attack-related event**). If the predecessor event has multiple occurrences (in this case, every year from year 1 through year 75) but the event occurs only after the *last* occurrence, then checkmark the **Only after last occurrence** check box. If you want the event to occur the same year that the predecessor does, then set the **Lag** to zero; if, on the other hand you want the event to have some lag, then set the **Lag** to the years between these events.

Depending on your settings, a **Concrete frequency model** list box may also be available. You can use this selection to set the lag based on the predictions of the BridgeLCC Concrete Service Life Prediction Tool. (See Section 3.3 for details on its use.)

**Event timing**

The setting of the timing of an event is identical to that in the **Edit Costs** window, with one important exception. As with cost timing, an event that occurs only once is set by leaving the **Repeating** box unchecked and selecting a year from the **Start year** list box. If the event occurs more than once, then checkmark the **Repeating** box and select values from the **Start year** and **End year** boxes.

If, as in the case of **No attack**, the event occurs after another event, then the **Start year** text changes to **First year after**, and you input *the number of years after the predecessor occurs that the current event should first occur*. If the event is recurring, then input the last year *after the predecessor* this event should occur. For example, if the last occurrence of the predecessor were year 25, and you want your event to occur in years 30 through 35, then input "5" in the **First year after** box, "10" in the **Last year after** box, and "1" in the **Recurs every ___ years** box.

As with the **Lag** box, a **Concrete frequency model** list box, when available, can be used to set the frequency of occurrences. For example, if the frequency of your event occurring reflects a recurring deterioration/repair cycle, which is itself modeled through the BridgeLCC Concrete Service Life Prediction Tool, then select one of the mix designs listed in the **Concrete mix** list; BridgeLCC will then automatically set the **Recurs every ___ years** field. See Section 3.3 for more details on using the BridgeLCC Concrete Service Life Prediction Tool.

**Event probability**

Events can also be modeled as probabilistic events, such as the occurrence of earthquakes or terrorist strikes. For example, if the event were a Richter-5 earthquake, then set the probability to some reasonable value, such as "0.05" (percent). The **Other probabilistic successors** box lists the other events that also follow this event's predecessor.

To confirm your data, the lower-left graph displays the years in which the currently displayed event will occur, and the right-side panel shows the current structure of all events for this alternative.

30

## 3.2    Workzones

The most common way of measuring the impact of bridge construction and repair activities on the drivers on and under the bridge are through the calculation of *user costs*. These are often measured as the sum of (1) the additional costs to drivers from the delays that bridge activities cause, or *driver delay costs*; (2) the additional costs to companies whose vehicles are delayed, or *vehicle operating costs (VOCs)*; and (3) the costs from the additional frequency of accidents in the bridge workzone, or *accident costs*.

BridgeLCC provides the **Workzones** tab in the **Project Assumptions** window (Figure 25) for managing a database of traffic-related user cost. (The formulas used for computing the driver delay, vehicle operating, and accident costs are described in Appendix B.)

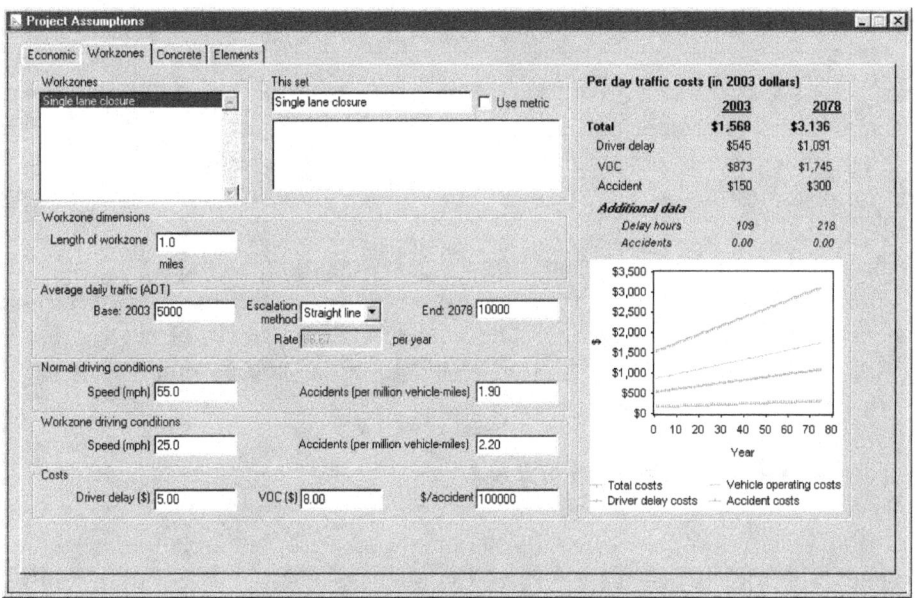

**Figure 25. Workzones Tab**

As with the **Edit Costs** and **Edit Events** windows, a set of workzone data can be created or deleted by positioning the mouse over the **Workzones** list box and clicking the right mouse button; a "pop-up" menu will appear listing these options.

The lower-left panels list the categories of data necessary for computing these costs.

**Workzone dimensions**

BridgeLCC computes driver delay and vehicle operating costs as a function of the additional time a driver spends over the length of the workzone, specifically where the driver moves at a reduced speed. Input this length in the **Length of workzone** field.

**Average daily traffic (ADT)**

Workzone costs increase as the number of vehicles that pass through the zone increases; this number is typically expressed as the average daily traffic (ADT). The **Base** and **End** fields list the traffic levels at the beginning and end *of the study period*. Since traffic typically increases over time, BridgeLCC provides two means of increase: **Straight line** (linear) and **Exponential**. When **Straight line** is selected, the **End** field is available; when **Exponential** is selected, the **End** field is disabled but the **Rate** field becomes available. Input your appropriate values.

**Normal driving conditions, Workzone driving conditions**

For the **Normal driving conditions** fields, input the average driving speed and accident rates when there are no construction activities; for the **Workzone driving conditions** fields, input the speed and accident rates during workzone activities (such as deck repairs). The differences between these normal and workzone rates are used to compute the additional time spent by each driver in the workzone.

**Costs**

Input the hourly costs to drivers and to vehicle owners of spending each additional hour driving and the cost of each additional accident. In Figure 25 the cost per hour of driver delay is $5.50 per hour and of vehicle operating cost is $11.00 per hour; the cost per accident is $200,000, *all expressed in base year dollars*.

The upper-right panel summarizes the user costs for this workzone, including the breakdown by type of user cost, the delay hours per day (the delay per car times the number of vehicles), and the increase in accidents per day. It also graphs the breakdown and growth of user costs over the study period. (See Appendix B for a full description of the formulas and assumptions used in these user cost calculations.)

**3.3    Concrete Service Life Prediction Tool**

BridgeLCC also includes a tool for estimating the time it takes for concrete to deteriorate to the point of needing repair. The tool is based on research conducted at NIST's Building and Fire Research Laboratory; see http://www.bfrl.nist.gov/862/vcctl/ for specific details on this research.

This service life tool is integrated into a BridgeLCC analysis the following way. First, the user designs alternative concrete mixes in this **Concrete** tab in the **Project Assumptions** window; each results in an estimated time-to-repair of the particular concrete structural element. Next, in the **Edit Events** window, the user selects the mix designs as estimates of either the lag between sequential (i.e., predecessor-successor) events (in the **Lag** field) or the frequency with which an event repeats (the **Recurs every ___ years** field).

Figure 26 illustrates the components in this window.

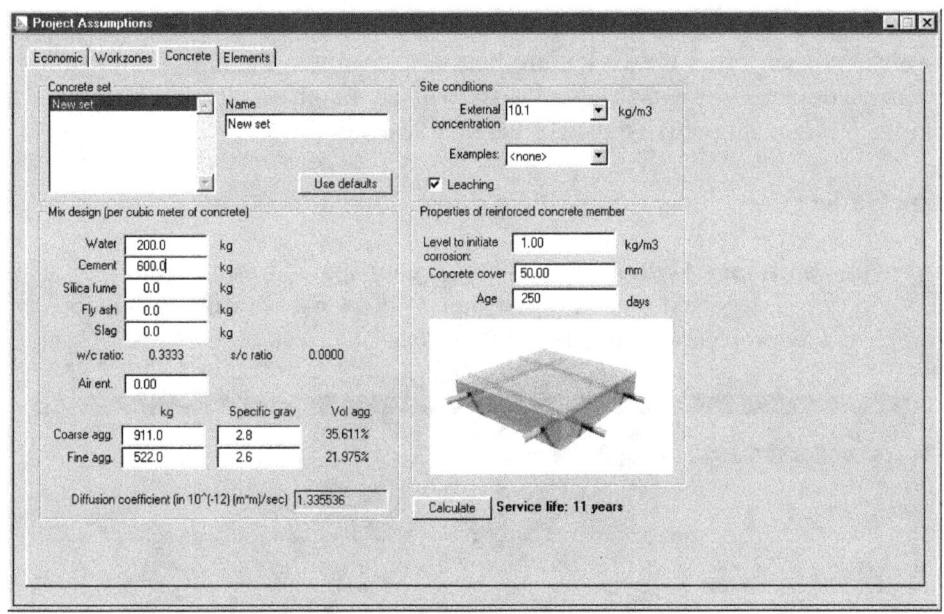

**Figure 26. Concrete Service Life Prediction Tool**

### Concrete set

New concrete sets can be created by positioning the mouse over the list box, pressing the right mouse button, and then selecting **Add new item** from the pop-up menu. Press the **Use defaults** button to reset all values to a default set provided by BridgeLCC.

### Mix design

Use the fields in this panel to define the mix design of our concrete. The panel provides useful mix metrics, including the water-to-cement ratio (**w/c ratio**), the silica-to-cement ratio (**s/c ratio**), and the sum of the volumes of all ingredients, which should sum to 1 cubic meter. The **Diffusion coefficient** field lists the chloride diffusion rate for this mix in $1 \times 10^{-12}$ meters-squared per second.

### External site conditions

The two site conditions that affect the rate of chloride diffusion are (1) the external exposure to chlorides (road salts) and (2) whether leaching occurs. Either input a level of exposure in the **External concentration** field or select a representative state from the **Examples** list box. Checkmark the **Leaching** box if leaching occurs (leaching increases the rate of diffusion).

### Internal conditions

The three internal factors affecting the initiation of corrosion are (1) the minimum internal level of chlorides necessary to start the corrosion of internal reinforcing steel, (2) the length of concrete cover (i.e., the distance over which the chlorides must travel from the outside to the reinforcing steel), and (3) the number of days between when the concrete was first poured and the first contact with external chlorides occurs. Input these three respective values in the **Level to initiate corrosion**, **Concrete cover**, and **Age** fields.

Once these values are set, press the **Calculate** button to compute the chloride diffusion coefficient and service life of this mix design. (These values will be automatically computed when life-cycle costs are computed for an alternative that uses the mix design.)

## 3.4    Image Gallery

BridgeLCC provides an **Image Gallery** window for organizing and printing images relating to your analysis. The window is accessed via the left panel in the **Cost Summary** window or the menu. Images are organized by project, alternatives, and results.

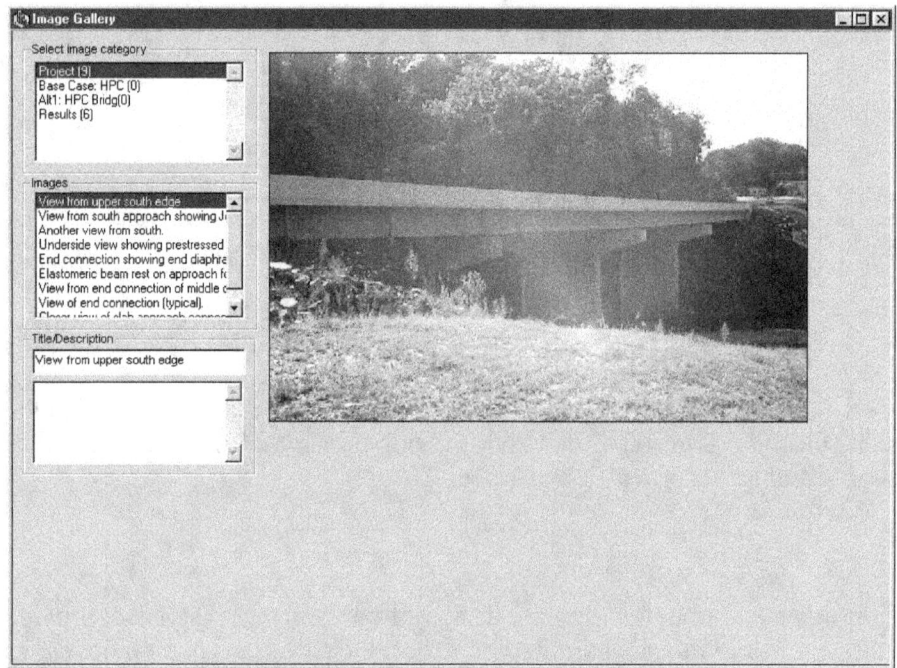

**Figure 27. Image Gallery Window**

To input a new photographic or line art image, select a **Category** (**Project**, an alternative, or **Results**) by positioning the mouse over the current list of images for that category and pressing the right mouse button; a menu will appear, listing options for adding a new image (or deleting the currently selected image). If a displayed image appears smaller than its original size, then try enlarging the **Image Gallery** window as much as possible; the image should resize to the window.

Most of the life-cycle costing graphs in BridgeLCC can also be added to the **Image Gallery**. For example, snapshots of the cost summary and cost timeline graphs can be added to the **Image Gallery** by positioning the mouse over these graphs, right clicking the mouse button, and selecting **Add graph to gallery** from the pop-up menu.

## 3.5    Online Help

BridgeLCC provides context-sensitive help for all of its windows. To access this help for the current window, press the **F1** key. Press **F6** to access the Table of Contents, shown in Figure 28. The help also provides a **Search** tab, which is used to search the help for specific terms, such as "inflation rate."

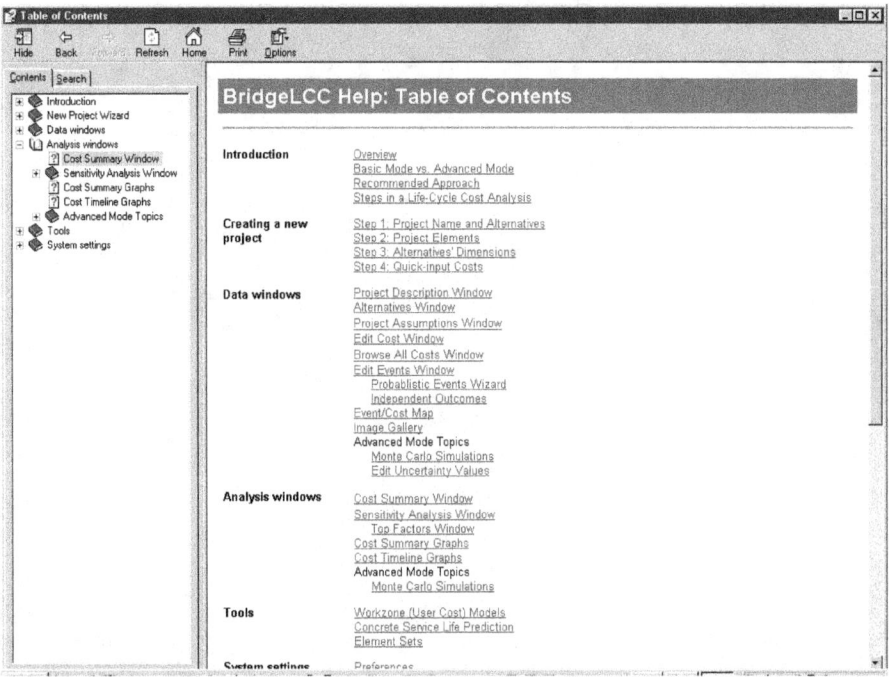

**Figure 28. Online Help**

## 3.6    Preferences Window

The **Preferences** window (Figure 29) is used to set the defaults for particular optional windows, graph colors and fonts, inflation and real discount rates, and analysis modes (either Basic or Advanced). Once the selections are made and the window closed, these settings will be used in all BridgeLCC analyses.

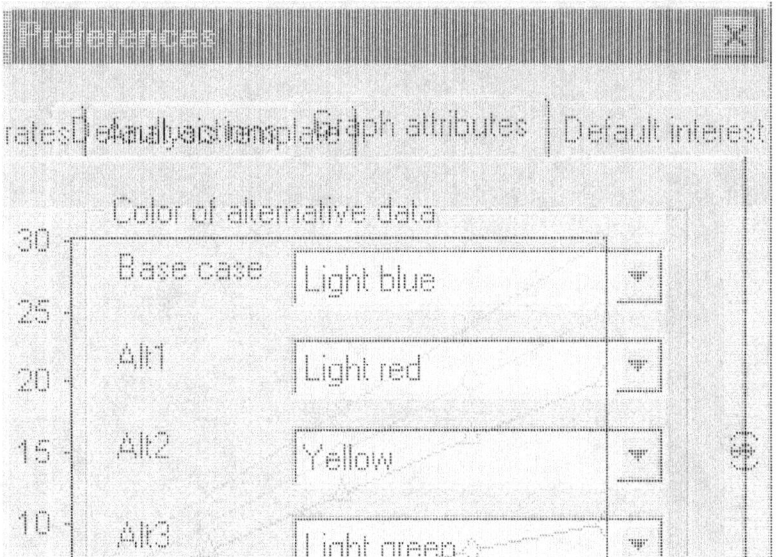

**Figure 29. Preferences Window**

35

# 4. Sensitivity Analysis

In the Basic Mode, all of the values used for costs, real discount rate, inflation rate, and others are considered to be best-guess values; they provide a deterministic estimate of the life-cycle costs of each alternative. Still, the user would like to know the influence of each variable on the life-cycle costs and cost competitiveness of the alternative —would the alternative no longer be cost competitive if the variable increased or decreased slightly?

In lieu of conducting a comprehensive analysis of uncertainties in the Advanced Mode, the user can analyze the effect of individual parameters on life-cycle costs. This can be achieved with three tabs in the **Sensitivity Analysis** window:

1. The **Change in a Single Factor** tab,
2. The **Most Significant Factors** tab, and
3. The **LCC Snapshots** tab.

## 4.1    Change in a Single Factor Tab

Consider a two-alternative BridgeLCC analysis, in which the deterministic best-guess values show Alternative 1 to be the life-cycle cost-effective design. A common question is: is this alternative the cost-effective design when the real discount rate changes?

The **Change in a Single Factor** tab can be used to answer this question. As shown in Figure 30, the user can select **Discount Rate** from the left panel, select +/- **100%** in the **Variation** box, and then press the **Compute** button; the graph will then display the life-cycle costs of each alternative over the range of real discount rate values.

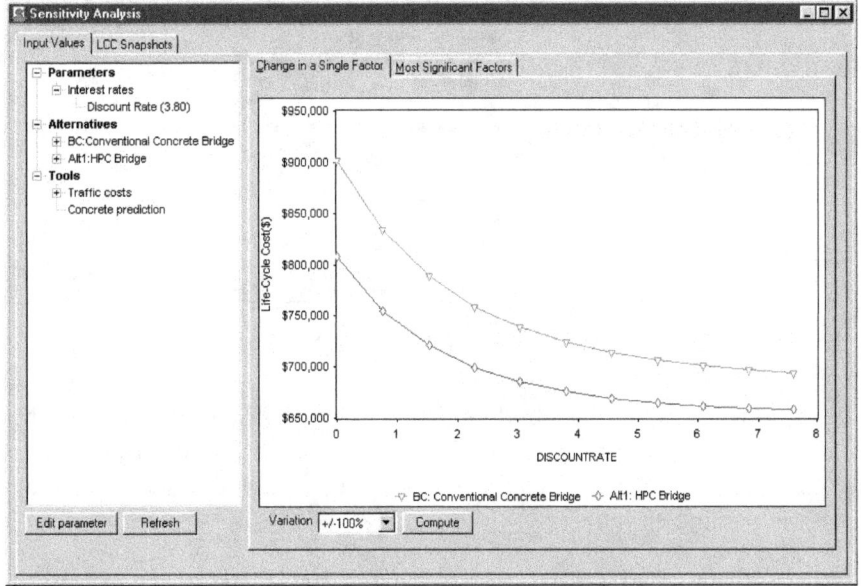

**Figure 30. Change in a Single Factor Tab**

As shown, Alternative 1 has lower life-cycle costs than the Base Case for real discount rates between 0.0% and 7.6%. Said another way, Alternative 1 is the life-cycle cost-effective alternative regardless of the discount rate used.

The left-side panel lists all costs, events, and other parameters used in some or all of the alternatives in the analysis. Select an item to see its effect on life-cycle costs and whether one alternative remains the best choice in terms of life-cycle cost.

To print the graph, add it to the **Image Gallery** window (for inclusion in analysis reports), or to copy it for pasting into a word processing or presentation file, position the mouse over the graph and press the right mouse button; a window will appear with these options.

## 4.2 Most Significant Factors Tab

As regular use of the **Change in a Single Factor** tab will show, some factors have more impact on life-cycle costs than others. Those factors that have little or no effect on life-cycle costs can in fact be ignored. The **Most Significant Factors** tab (Figure 31) is used to rank-order the effect of parameters on life-cycle costs so that the most important factors can be identified and investigated further.

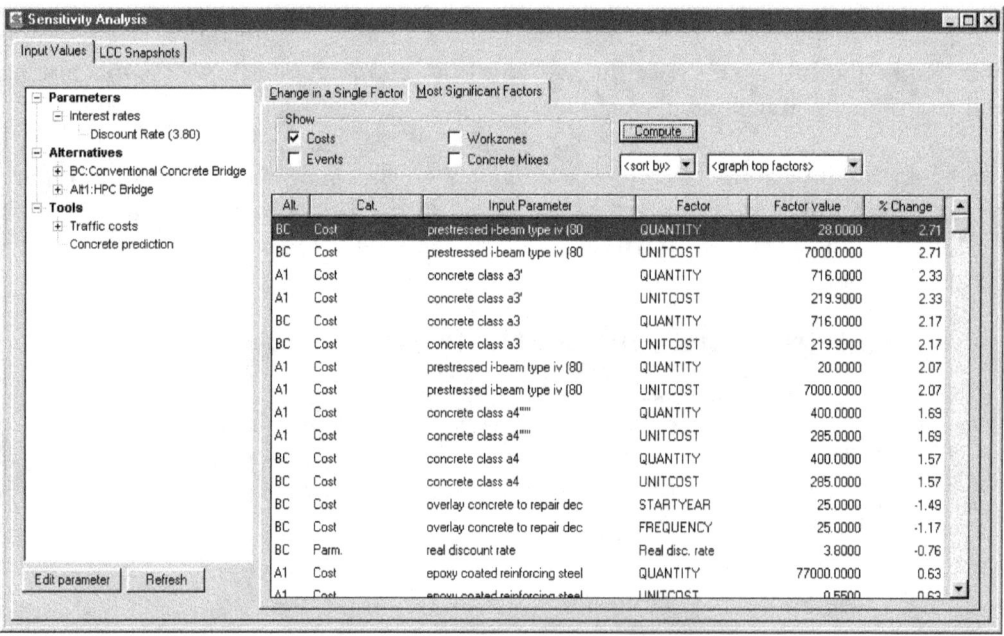

**Figure 31. Most Significant Factors Tab**

To compile this list, press the **Compute** button. The window will then compute, for each parameter listed in the left-side panel, life-cycle costs based on a 10% increase in the value. The percent change column lists the percent change in life-cycle costs (for each alternative that uses the parameter) due to the 10% increase in parameter value.

When done, the window will list the most important factors, in descending order. To view a graph of these factors (Figure 32), select a number from the **<graph top factors>** box. To print this graph, add it to the **Image Gallery**, or copy and paste it to a word-processing or presentation document, position the mouse over the graph and click the right mouse button; a menu of choices will appear.

38

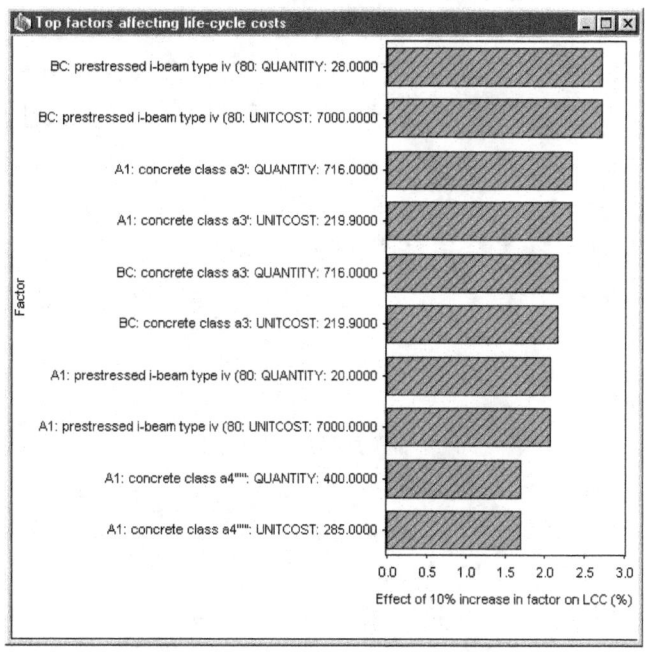

**Figure 32. Top Factors Graph**

## 4.3    LCC Snapshots Tab

Analysis of single and multiple factors may lead the analyst to want to compare the baseline set of best-guess values with another set of values. The summary values of these two or more sets of values can be compared using the **LCC Snapshots** tab, shown in Figure 33.

The simple case works as follows: first, after completing the input of all best-guess values for all alternatives, save the analysis, and then create a snapshot of these values. Take a snapshot by (1) creating a new snapshot entry (right-click the mouse over the list of snapshots; a menu will appear for adding and deleting snapshots) and (2) pressing the **Take snapshot** button. The cost fields in the lower panel should now show the values currently shown in the **Cost Summary** window.

Next, change the values of costs, events, and other parameters to reflect a coordinated set of alternative values (such as changing both inflation and real discount rates). Take a second snapshot, using the two steps above. When done, print both snapshots (through the **Reports** window) and compare.

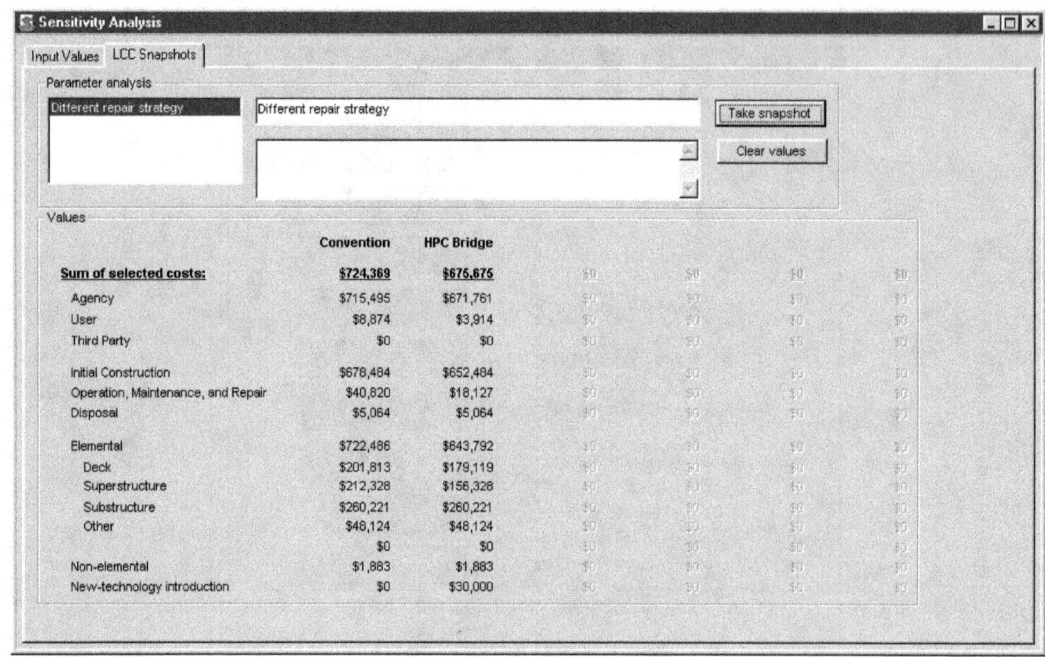

**Figure 33. LCC Snapshots Tab**

# 5. Uncertainty and Risk Analysis (Monte Carlo Simulations)

## 5.1 The Basic and Advanced Modes

In some cases, the best-guess values are not known with significant certainty and should rather be viewed as ranges of values. These ranges of costs, timings of costs, events, discount rates, and other parameters result in ranges of life-cycle costs. Each alternative does not have a single life-cycle cost but rather a range of life-cycle costs.

BridgeLCC separates out uncertainty and risk analysis by providing two modes of use, the latter of which provides fields for inputting and analyzing uncertainty. In the first mode, called the *Basic Mode*, there are no fields for inputting uncertainty, other than inputting in the **Edit Events** window the probability of an event occurring. In the second mode, called the *Advanced Mode*, there are uncertainty fields associated with many variables. The user can switch back and forth between these two modes without any loss of data. As an example comparison, Figure 34 shows the **Edit Costs** window in both modes; the *Advanced Mode* version has additional fields for inputting uncertainty values for **Start year**, **End year**, **Every ____ years**, **Quantity**, and **Unit cost**.

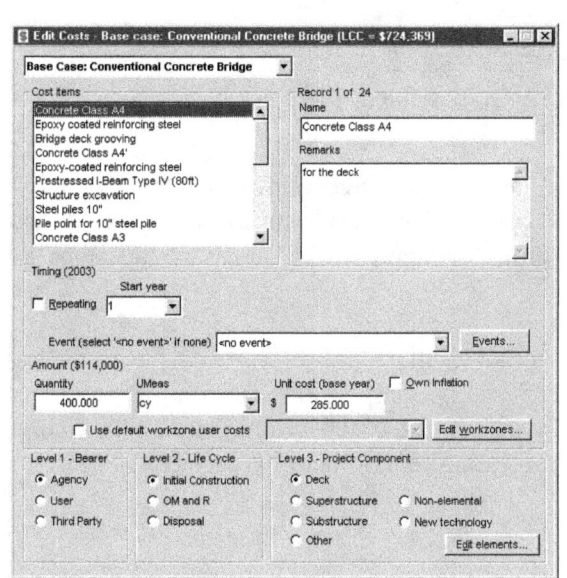 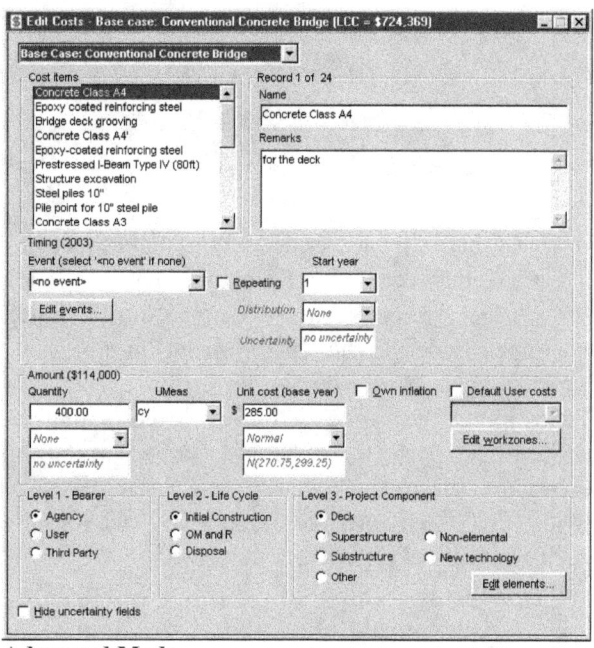

Basic Mode                                        Advanced Mode

**Figure 34. Basic and Advanced Mode Versions of Edit Costs Window**

To move between the two modes, either press the **Go Advanced** or **Go Basic** button in the **Cost Summary** window. To make the current mode the default mode for all BridgeLCC analyses, press the **Set as default** button in the **Cost Summary** window.

Since using a second, advanced mode adds a layer of complexity to the overall process, it is recommended that you use the following steps to conduct an uncertainty and risk analysis in BridgeLCC:

1.  First, in the BridgeLCC *Basic Mode*, input all of the best-guess values for all costs, events, and other parameters in all alternatives. Once complete, note the life-cycle cost of each alternative using the **Cost Summary** window.

2.  Switch to the BridgeLCC *Advanced Mode*. For each parameter to be modeled with a range of values, select a distribution type from the parameter's **Distribution** list box. The **Edit Uncertainty Values** window will appear (Figure 35), allowing for the setting of the values needed for that particular probability density function.

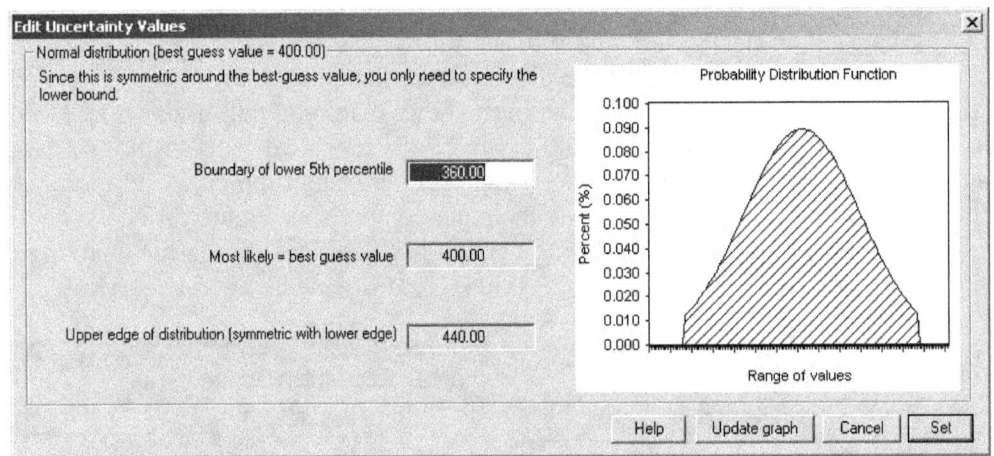

Figure 35. Edit Uncertainty Values Window

3.  Once all uncertainty values have been inputted, access the **Uncertainty and Risk** window (either by clicking **Uncertainty** on the left-side panel in the **Cost Summary** window or via the **Analysis/Monte Carlo Simulation...** in program menu choice). Follow the steps as provided in the **Run Simulation** tab.

To estimate the range of life-cycle costs that can result from the range of values of parameters, BridgeLCC performs a Monte Carlo simulation: it repeatedly samples the uncertain values and computes life-cycle costs for all alternatives. Since each of these repetitions is equally likely, BridgeLCC then sorts the outcome by size and plots it, resulting in a probability density function for each alternative's life-cycle cost.

## 5.2    Run Simulation Tab

Monte Carlo simulations are started by selecting the number samples to take and the set to which the results are to be stored. Input these values in the fields in the **Run Simulation** tab, shown in Figure 36, and then press the **Run** button.

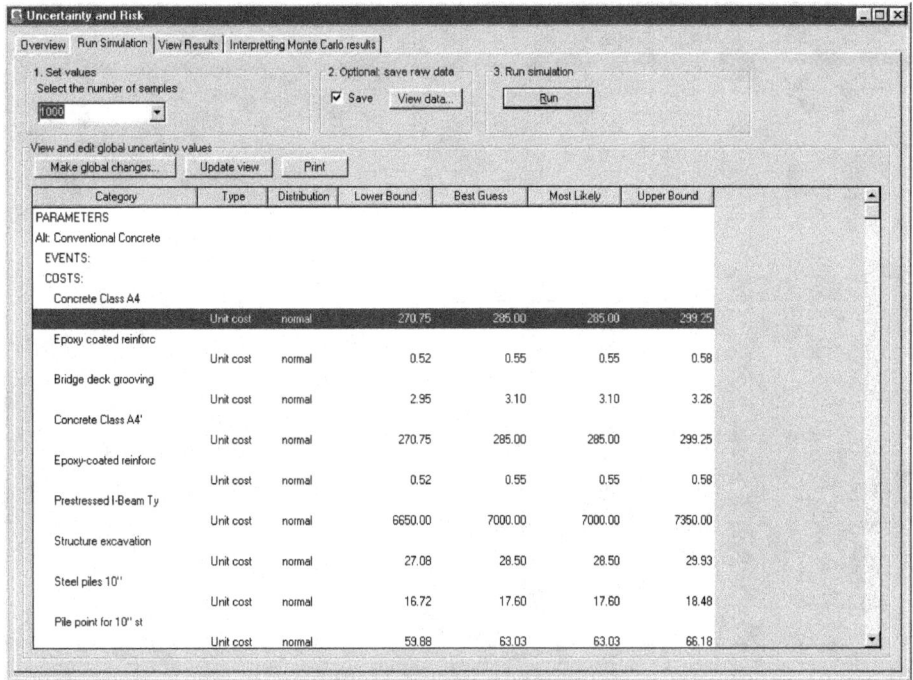

**Figure 36. Run Simulation Tab**

During the simulation, the window will display the number of samples taken thus far and the time remaining to completion. When the simulation is completed, access the **View results** tab. Also shown is a listing of all the uncertainty values for *only* the alternatives and cost types currently check-marked in the **Cost Summary** window.

## 5.3    Raw Data

In addition to creating a summary graph and set of statistics, BridgeLCC can display the raw results of the simulation. Since creating this data file can increase the time it takes to run the simulation, you must first checkmark the **Save** box in the **Run Simulation** tab. When the simulation completes, press the **View data ...** button that appears after the simulation is completed.

## 5.4    View Results Tab

The **View results** tab (Figure 37) displays the distribution for each alternative simulated, as well as summary comparative statistics. To create the graph, BridgeLCC computes the difference between the lowest and highest life-cycle costs from all alternatives and then creates 20 equal size "bins." The life-cycle cost from each alternative is assigned to a bin, and then the count in each bin is normalized so that the set of 20 bins represents the density function of each alternative.

43

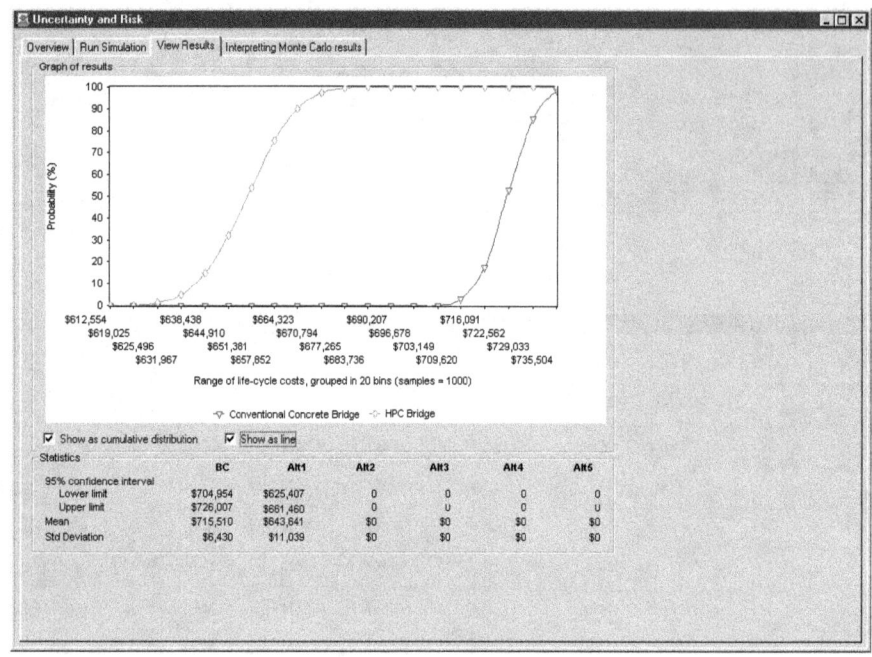

**Figure 37. View Results**

The lower **Statistics** panel displays useful statistics for each alternative. The **95% confidence interval** values are the range over which 95% of the samples were observed; for data that is normally distributed and therefore has no statistical lower or upper bounds, this interval is a useful measure of where "most" observations lie. The **Mean** is the average value of each alternative's samples, and will likely be very similar to the best-guess values displayed in the **Cost Summary** window. The **Standard Deviation** is another measure of the data's variation.

To show the probability density function as a line, checkmark the **Show as line** box. To show the cumulative density function (the cumulative probability density), checkmark the **Show as cumulative distribution** box.

## 5.5    Global Uncertainty Changes Window

To make easier the task of changing the uncertainty values used throughout the cost, event, and parameter windows, the **Global Changes** window (Figure 38), accessed via the **Make global changes ...** button in the **Run Simulation** tab, displays all of the current uncertainty settings, and contains fields for making global changes to uncertainties.

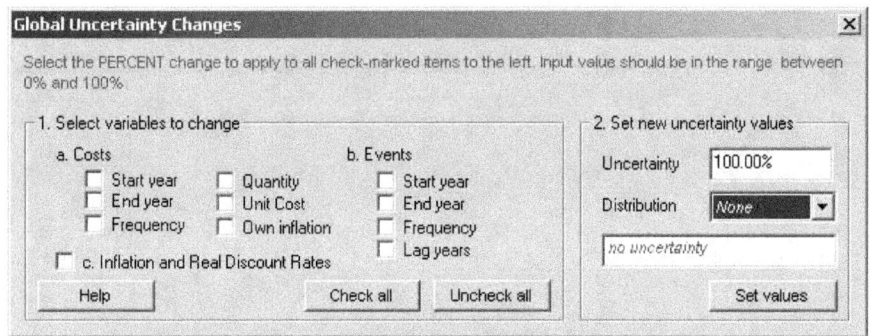

**Figure 38. Global Uncertainty Changes**

To make global changes to the uncertainty settings:

1.  Access the **Cost Summary** window and checkmark both the alternatives and cost types whose uncertainty values need to be changed.

2.  Access the **Global Uncertainty Changes** window via the **Make global changes...** button in the **Run Simulation** tab. Select the types of values to be changed in the **Select variables to change** group, and then select the distribution type in the **Distribution** list box. The **Edit Uncertainty Values** window (Figure 35) will appear, allowing you to set the parameters for this uncertainty.

3.  Press **Set** in the **Edit Uncertainty Values** window to set the uncertainty, expressed in percent terms (since you may be simultaneously setting uncertainty for dollar, quantity, and time amounts). This will also close the **Edit Uncertainty Values** window.

4.  Press **Set Values** in the **Global Uncertainty Changes** window to set this percentage uncertainty to the selected parameters.

## 6. Example Analysis – Basic: Route 40 HPC Bridge

This chapter provides an example life-cycle cost analysis to illustrate in more detail how BridgeLCC is used to assess competing, alternative construction materials. The level of detail shown is meant to be typical. More or less detail can be used depending on the needs of the engineer and availability of data. This analysis is saved as the file "Route40.lcc," located in your BridgeLCC directory.

In this example, an engineer is making a preliminary design of a highway bridge and is considering two alternative types of concrete. The base case concrete is the conventional mix currently used by the engineer. The alternative concrete mix is a high-performance concrete (HPC) that the engineer has not used before, but that should produce stronger and more durable bridge components. The engineer wants to determine which of the two materials is life-cycle cost effective for this bridge.

The engineer follows the BridgeLCC steps, which are based on ASTM practice E 917 for measuring the life-cycle costs of buildings and building systems (see Appendix A for a description of the method).

This chapter is divided into three sections, each covering a logical division of this group of steps. Section 6.1 describes the project objective on which the life-cycle cost analysis is based, including the bridge performance requirements and the material alternatives that meet those requirements. Section 6.2 describes the project parameters which are independent of each alternative, such as the daily traffic volume on the bridge, the number of years the bridge is required to last, and the interest rate at which future costs are discounted to the present.

Next, the best-guess costs for the conventional concrete and the high-performance concrete are estimated, starting first with the costs incurred by the agency (the department of transportation [DOT]). Following ASTM E 917, these costs are divided into initial construction costs; operation, maintenance, and repair (OM&R) costs; and disposal costs. Each of these three are further divided into groups based on the component of the structure to which the cost is tied. For example, the agency initial construction costs (i.e., the engineer's estimate) are divided into deck, superstructure, substructure, "other," non-elemental, and new-technology introduction costs.

Section 6.3 describes how to interpret the computed best-guess life-cycle costs of each bridge design alternative. The life-cycle costs of each are compared using graphs that display breakdowns according to BridgeLCC's cost classification scheme. The engineer then revisits each alternative's best-guess costs and determines to what extent uncertainty in these costs affects the overall life-cycle costs of each alternative bridge. The chapter ends with a summary of the sample analysis.

The following sections are written from the perspective of an engineer who is making a preliminary design of a bridge and, as part of that process, comparing the life-cycle costs of two competing, alternative materials: conventional-strength concrete and high-performance concrete.

## 6.1　Overview

**Step 1:**
**Define the project objective and minimum performance requirements.**

The project objective is to build, maintain, and eventually dispose of a new highway bridge in a specific rural Virginia county. The engineer first makes a general description of the size of the bridge and the environment in which it will exist. The structure is shown in Figure 39. It is 100 meters (322 ft) long, 14.5 meters (47 feet) wide, and will carry two lanes of traffic over a stream. The bridge is part of a rural highway that has relatively low volumes of traffic. Winter precipitation and temperatures are mild, with only 5-10 days on average that require snow plowing and salting of roads each year.

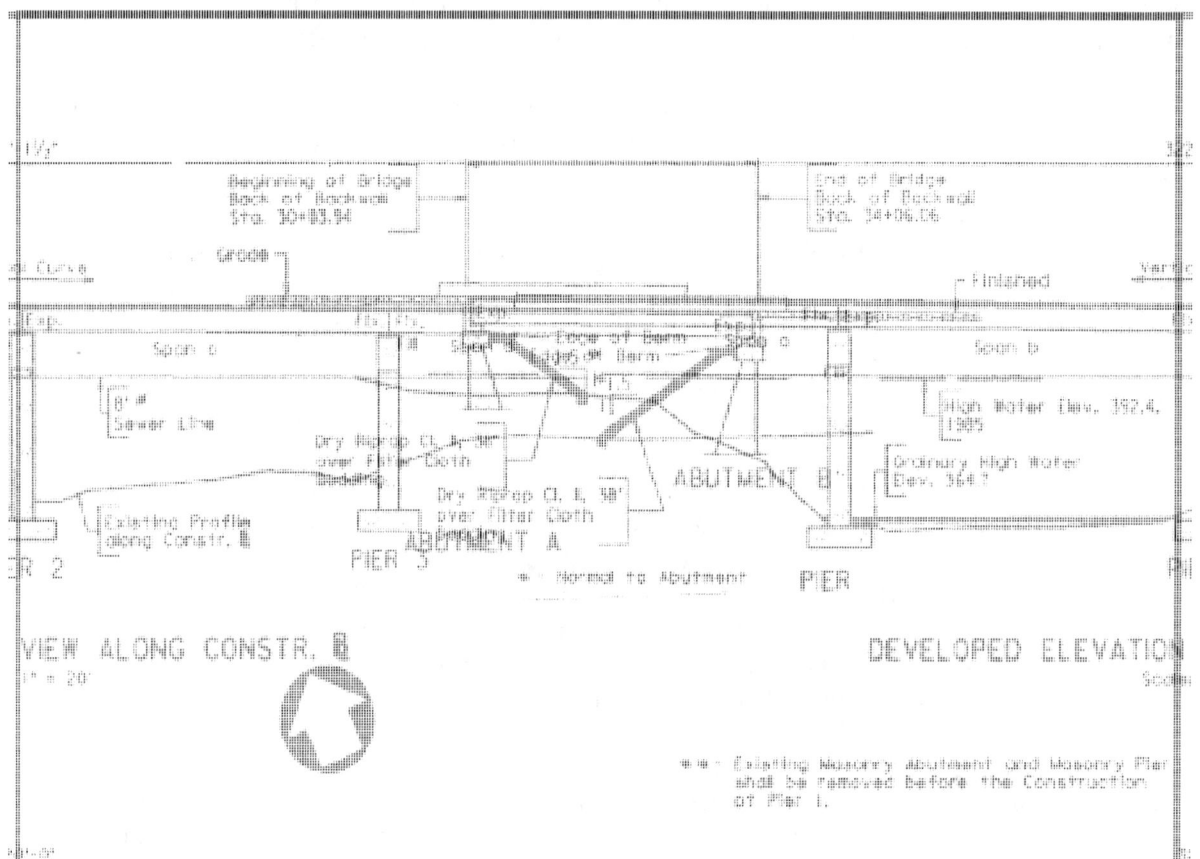

**Figure 39. Plan and Elevation Views of Bridge**

The engineer next lists the minimum performance requirements of the structure. Both the conventional concrete bridge and HPC bridge must satisfy these performance-based requirements. The structure must be able to carry the loads prescribed in AASHTO's[5] HS-20 specification. The spans between piers must not deflect more that L/800 meters, where L is the distance between the piers. The bridge has a design life of 75 years.

---

[5] American Association of State Highway and Transportation Officials.

Table 1 lists the technical characteristics of the two alternative concretes that meet the above performance-based requirements. Sufficient detail is listed so that the differences in total life-cycle costs can be understood in terms of these technical differences. For example, if it turns out that one bridge deck's repair cost is 50% less than the other, the engineer can ascertain that it is because the concrete is twice as durable.

**Table 1. Technical Characteristics of the Base Case and Alternative Bridge Designs**

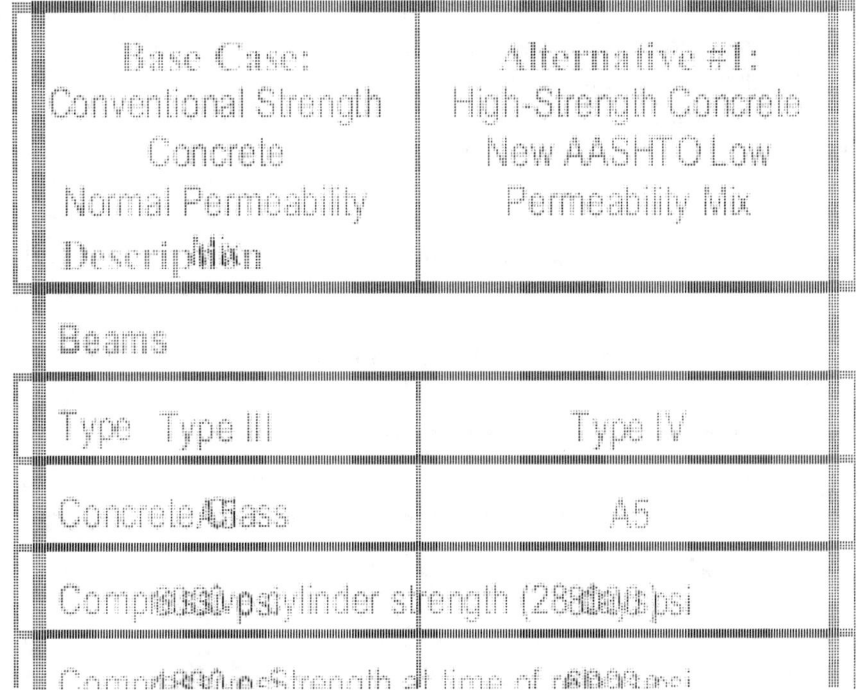

To estimate costs, the engineer needs to carefully define how the bridge will be built, maintained and repaired, and disposed of. The base-case, conventional-concrete bridge is built by first driving piles into the stream bed, pouring footings around the piles, forming and pouring the piers and bents, installing precast prestressed beams, and then pouring the deck and approaches in place. Every 25 years the deck will be repaired for environmental corrosion and mechanical wearing; after 75 years the bridge will be demolished.

The alternative, HPC bridge uses high-strength concrete in the deck and beams. In addition, the deck uses an AASHTO mix design that reduces the permeability of the concrete, thereby reducing the intrusion of road salt chlorides and ultimate corrosion of the reinforcing steel. The HPC bridge is built, maintained, and disposed of the same way as the base case bridge but with two important exceptions. First, because of its higher-strength concrete beams, the alternative bridge has five beams between piers instead of seven. Second, because of the AASHTO low-permeability specification, the deck requires repair every 40 years instead of every 25.

## 6.2   Data

### Project Parameters

Step 3:
Establish the basic
assumptions for
the analysis.

The project parameters quantify the conditions the bridge will experience regardless of the concrete used. The engineer starts by itemizing the factors that will affect the bridge over its life cycle. Three common factors are traffic conditions, weather conditions, and economic conditions.

Traffic conditions, such as average daily traffic (ADT) and accident rates, determine the costs that bridge construction, repair, and disposal place on drivers who travel over the bridge (there is a stream under the bridge, so these costs are ignored for traffic below the bridge). Weather conditions determine how much road salt is placed on the road. Economic conditions, specifically the inflation rate and real discount rate (which are used in the life-cycle costing formulas), determine the relative importance of costs that occur later in the life of the structure.

Table 2 lists the traffic data the engineer compiles for the Virginia bridge. Departments of transportation often have forecasts of the traffic levels a bridge is expected to experience.

### Table 2. Project Parameters

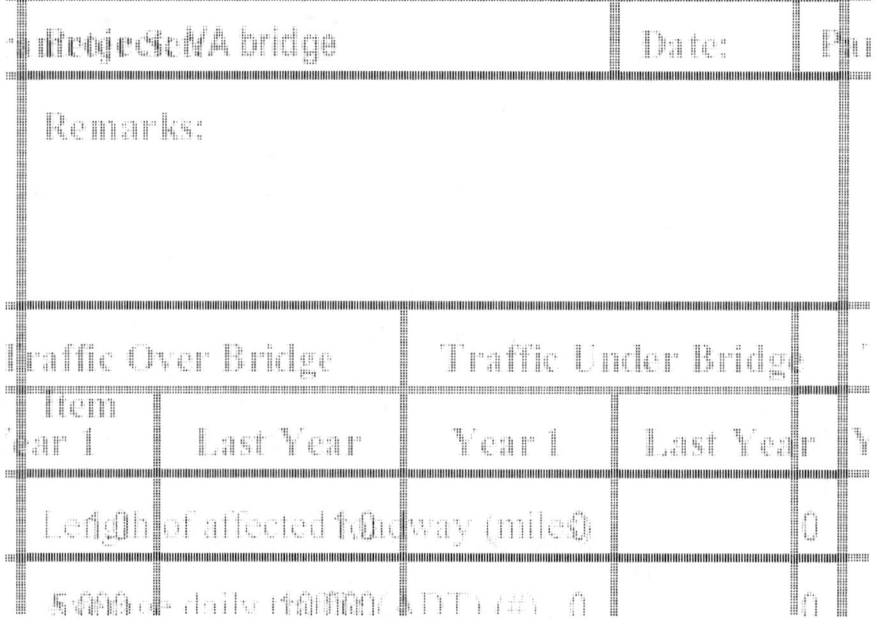

The "Length of affected roadway" is the length of road over which traffic is slowed or diverted. "Average daily traffic (ADT)" is the average number of cars that drive over the bridge every day; the engineer estimates that this traffic volume will increase from 5,000 to 10,000 over the 75-year life of the bridge.

"Normal driving speed" is the traffic speed when there is no bridge work affecting traffic flow, while "Roadwork traffic speed" is the speed of traffic when such traffic is diverted around roadwork. The

"Normal accident rate" and "Roadwork accident rate" are estimates of how often accidents occur outside of and during bridge work (the rates are in accidents per million vehicle miles, where vehicle miles is the product of the number of vehicles and the average miles each car drives). The "Cost per accident" is the average total cost of an accident in the base year of the study period. "Hourly driver cost" is the value of drivers' time, and "Vehicle operating cost" is the value of commercial vehicles' time in the base year of the study period. If the engineer does not have one or more of the required values in-house, national values or values from other states can be used.

For the base case bridge, an engineer will likely know how often and how much the bridge will need repair in order to last 75 years (unless the engineer typically designs a bridge to last 40 years; in this case an explicit 75-year schedule may not be available). However, the engineer may not know how long the high-performance concrete bridge deck will last before repair is required. The level of road salt applied is the key parameter for determining how long the bridge deck will last. Table 3 lists some example values of road salt levels for different states.

**Table 3. Salt Exposure in Various States**

| Salt Level | Level in kg/m3 | States using this level |
|---|---|---|
| Low | 5.06 | KS, CA |
| Medium | 10.11 | MN, FL |
| High | 15.17 | DE, IA |
| Severe | 20.90 | WI, NY |

The engineer can use the **Concrete** tab in the **Project Assumptions** window to estimate and compare the service lives of the two alternatives and to generate repair schedules for them. The engineer can later change the repair schedules to see how sensitive the life-cycle cost of each alternative is to repair costs. If the costs are not sensitive to changes, then the engineer can conclude that the life-cycle costs are robust to changes in the repair schedules.

The last design-independent project parameters are the inflation rate and real discount rate. The inflation rate is the average rate at which prices will increase over a given year. The real discount rate is the rate at which future costs are discounted to base-year dollars. Federal infrastructure projects use a discount rate published in OMB Circular No. A-94. Table 4, reprinted from the circular, shows what the real discount rate is for different time horizons.

**Table 4. Real Discount Rate, by Number of Years**

| Number of Years | Discount Rate (%) |
|---|---|
| 7 years | 3.0 % |
| 30 years | 3.8 % |

Since the bridge has costs that occur as much as 75 years in the future, we use the maximum number of years in the table, 30, and the corresponding rate, 3.8%. Private-sector projects may use a different discount rate. The engineer inputs the traffic and economic parameters in the appropriate tabs in the **Project Assumptions** window.

**Project Alternative Cost Data**

| Step 4: Identify, estimate, and determine the costs |
|---|

The second type of data required by BridgeLCC is the set of costs for each bridge alternative. The engineer uses the **Cost Summary**, **Edit Costs**, and **Browse All Costs** windows and the cost classification scheme to systematically input the constituent costs. In this example the engineer first inputs the base case costs and then the Alternative #1 costs. In this particular analysis, the engineer starts with the agency costs and then inputs the user costs and third-party costs.

**Base Case Material: Conventional-Strength Concrete**

**Agency Costs**

The engineer organizes costs into initial construction costs; operation, maintenance, and repair (OM&R) costs; and disposal costs. Initial construction costs are a good place to start since many departments of transportation make an engineer's estimate of a bridge's construction cost before its drawings are sent out to bid. The engineer in this example uses the quantities of materials needed for the bridge and unit costs in an engineer's estimate from a previous structure to compile the initial construction costs. These costs (shown in Table 5) are then inputted into the **Edit Costs** window.

## Table 5. Example Cost Data

| Project: Route 40 in Virginia | | | | | | | | Date: | | | | pg# 1 | of | 3 pgs |
|---|---|---|---|---|---|---|---|---|---|---|---|---|---|---|

Remarks: Agency costs for the base case, conventional strength bridge. The HPC bridge will use the same costs, but reduce the number of beams and add new-material introduction costs.

| | | Cost Categories | | | Cost Quantities | | | | | | |
|---|---|---|---|---|---|---|---|---|---|---|---|
| Name | Remarks | Cost Bearer | Life-Cycle | Proj Comp | Qty | UMeas | Unit Cost | Range (±%) | Start Year | End Year | Freq |
| **INITIAL CONSTRUCTION COSTS** | | | | | | | | | | | |
| **DECK** | | | | | | | | | | | |
| Concrete | for the deck | agency | ic | deck | 400 | cy | $285.00 | 0 | 1 | 1 | 1 |
| Epoxy coated reinforcing steel | for the deck | agency | ic | deck | 77,000 | lbs | $0.55 | 0 | 1 | 1 | 1 |
| Bridge deck grooving | | agency | ic | deck | 1,566 | SY | $3.10 | 0 | 1 | 1 | 1 |
| **SUPERSTRUCTURE** | | | | | | | | | | | |
| Concrete | for diaphrams | agency | ic | super | 45 | cy | $285.00 | 0 | 1 | 1 | 1 |
| Epoxy coated reinforcing steel | for diaphrams | agency | ic | super | 6,000 | # | $0.55 | 0 | 1 | 1 | 1 |
| Prestressed I-Beam Type IV (80 ft.) | | agency | ic | super | 28 | ea | $7,000.00 | 0 | 1 | 1 | 1 |
| **SUBSTRUCTURE** | | | | | | | | | | | |
| Structure excavation | | agency | ic | sub | 702 | cy | $28.50 | 0 | 1 | 1 | 1 |
| Steel piles 10" | | agency | ic | sub | 1,190 | lf | $17.60 | 0 | 1 | 1 | 1 |
| Pile point for 10" steel pile | | agency | ic | sub | 40 | ea | $63.03 | 0 | 1 | 1 | 1 |
| Concrete Class A3 | | agency | ic | sub | 716 | cy | $219.90 | 0 | 1 | 1 | 1 |
| Reinforcing steel | | agency | ic | sub | 50,570 | # | $0.45 | 0 | 1 | 1 | 1 |
| Epoxy coated reinf. steel | | agency | ic | sub | 25,230 | # | $0.57 | 0 | 1 | 1 | 1 |
| | | | | | | | | | | | |

53

Table 5 (cont.)

| Project: Route 40 in Virginia | | | Date: | | | | | pg# 2 | of | 2 pgs |
|---|---|---|---|---|---|---|---|---|---|---|

Remarks: **Agency costs for the base case, conventional strength bridge. The HPC bridge will use the same costs, but reduce the number of beams and add new-material introduction costs.**

| Name | Remarks | Cost Categories | | | Cost Quantities | | | | | | |
|---|---|---|---|---|---|---|---|---|---|---|---|
| | | Cost Bearer | Life-Cycle | Proj Comp | Qty | UMeas | Unit Cost | Range (±%) | Start Year | End Year | Freq |
| Cofferdam | | agency | ic | sub | 2 | ea | $9,536.88 | 0 | 1 | 1 | 1 |
| OTHER CATEGORY | | | | | | | | | | | |
| Concrete | for parapet | agency | ic | other | 20 | cy | $285.21 | 0 | 1 | 1 | 1 |
| Epoxy coated reinf. steel | for parapet | agency | ic | other | 2,760 | # | $0.55 | 0 | 1 | 1 | 1 |
| Preformed Elastomeric Joint Sealer | | agency | ic | other | 252 | lf | $19.35 | 0 | 1 | 1 | 1 |
| Dry riprap | | agency | ic | other | 2,054 | tn | $17.49 | 0 | 1 | 1 | 1 |
| | | | | | | | | | | | |
| OM&R | | | | | | | | | | | |
| Overlay concrete to repair deck | | agency | omr | deck | 44 | cy | $1,200.00 | 0 | 25 | 50 | 25 |
| NBI inspection every two years | | agency | omr | non-el | 1 | ls | $150 | 0 | 2 | 74 | 2 |
| | | | | | | | | | | | |
| DISPOSAL | | | | | | | | | | | |
| Dispose of deck | based on cy concrete | agency | disp | deck | 0.33 | ls | $80,000 | 0 | 75 | 75 | 1 |
| Dispose of superstructure | based on cy concrete | agency | disp | super | 0.04 | ls | $80,000 | 0 | 75 | 75 | 1 |
| Dispose of substructure | based on cy concrete | agency | disp | sub | 0.61 | ls | $80,000 | 0 | 75 | 75 | 1 |
| Dispose of other | based on cy concrete | agency | disp | other | 0.02 | ls | $80,000 | 0 | 75 | 75 | 1 |
| | | | | | | | | | | | |

Following the cost classification, the engineer enters "agency" in the Level 1 (Cost Bearer) column, "ic" in the Level 2 (Life-Cycle) column, and the appropriate component in the Level 3 (Project Component) column.

The Start Year, End Year, and Frequency columns are used to enter the first year that the cost occurs, the last year the cost occurs, and the frequency with which the cost occurs. A frequency of "1" means that the cost occurs once a year, while a frequency of "5" means the cost occurs every five years. Since all of the initial construction costs for this bridge occur in the first year and only occur once, a "1" is inputted into the Start Year, End Year, and Frequency columns. All of the costs in the worksheet are inputted into the **Edit Costs** window.

Following the cost classification scheme, the engineer then generates a worksheet of all OM&R costs. Since the engineer estimates that there are no operation or maintenance expenses for the bridge, this is essentially a repair schedule for the structure. The engineer estimates that the base case bridge deck will require repair every 25 years. Each repair involves grinding off a thin layer of the bridge deck road surface and then applying a skim coat of new concrete. The engineer estimates this will take 40 cubic yards of concrete at a cost of $1,200 per cubic yard (which includes the cost of grinding off the old layer). No other maintenance or repair is required.

The final life-cycle category of cost is bridge disposal, which occurs in year 75. Since the cost is born by the agency and occurs during disposal, the Level 1 entry is "agency" and the Level 2 entry is "disp." The cost is broken down into disposal components based on the volume of concrete in each component. Since all costs occur in the last year of the life cycle and only occur once, the Start Year is "75," the End Year is "75," and the Frequency is "1."

All initial construction, OM&R, and disposal costs are now entered. The engineer can view and edit these costs in the **Browse All Costs** window. To view, for example, just the initial construction costs for the deck, the engineer can access the **Cost Summary** window, checkmark only the **Agency**, **Initial Construction**, and **Deck** check boxes (Figure 40), and then access the **Browse All Costs** window. The engineer can edit individual cost items in the **Browse All Costs** window by double-clicking the entry.

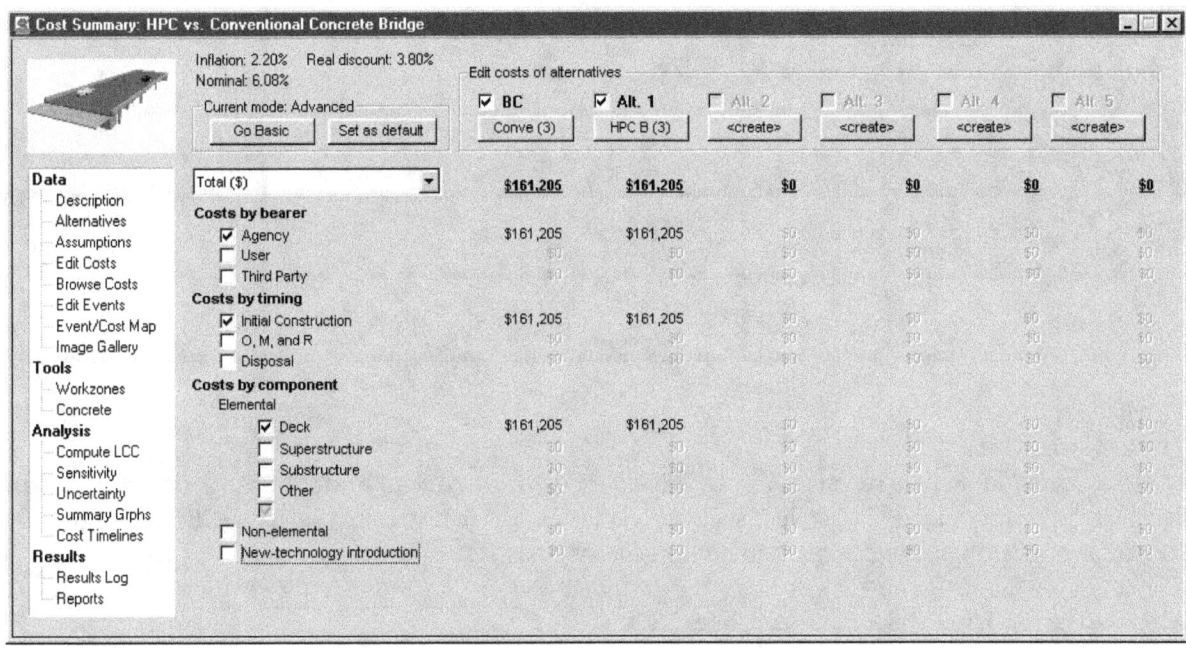

**Figure 40. Example Analysis - Basic: Filtering Costs Using the Cost Summary Window**

### User Costs

Traffic can be affected during construction, repair, and eventual disposal of the bridge. The *user costs* of these activities – the costs to users of the bridge – are estimated by creating new cost items in the **Edit Costs** window and then inputting the number of days that each bridge activity affects traffic. This data is used in conjunction with the traffic data in the **Workzones** tab in the **Project Assumptions** window to compute project user costs.

Table 6 lists the user costs for the two bridges.

**Table 6. Example User Costs**

| Project: Route 40 in Virginia | | | | | | pg# | of | pgs |
|---|---|---|---|---|---|---|---|---|
| Remarks: User costs for base case and alternative structure | | | | | | | | |

| Notes | Cost Quantities | | | | | Cost Category | |
|---|---|---|---|---|---|---|---|
| Proj Name | Qty | UMeas | Unit Cost | Range Remarks(±%) | Start Year | End Cost Year Range | Life Cycle |
| Base Case Bridge | | | | | | | |
| Redirect traffic during deck repair | | | year 25 and 50 | 50 | user | 25 | omit |
| | | | | | | | |
| Alternative Bridge (HPC) | | | | | | | |
| Redirect traffic during deck repair | | | year 40 | 40 | user | 1 | omit |

## Third-Party Costs

These costs could include revenues lost by adjacent businesses due to bridge construction, repair, and disposal, or environmental damage such as pollution of the stream under the bridge. For this particular project the engineer can find no third-party costs.

## Alternative #1: High-Performance Concrete

The engineer estimates that the costs of the alternative, high-performance concrete bridge are the same as the base case bridge, except for three important differences. First, the high-performance concrete allows the engineer to use 5 beams in each span instead of the 7 beams in the base case. This reduces the total cost of bridge beams. Secondly, the engineer requests that the beam fabricator perform some static load tests on one of the new-technology high-performance concrete beams to verify its load carrying capacity. This costs the DOT an additional $30,000. Finally, the new low permeability AASHTO specification for concrete extends the period between deck repairs from 25 years to 40 years. This reduces the number of repairs from 2 (in years 25 and 50) to 1 (in year 40).

The costs that differ from the base case bridge are show in Table 7.

**Table 7. Example HPC Costs that are Different than Conventional Concrete**

| Project: **Route 40 in Virginia** | | | | | | Date: | | | | pg# | of | pgs |

| Remarks: Costs in the HPC bridge design that differ from the base case design. The beam cost and repair costs below modify the values that were in the base case design. The static testing cost is a new cost item. |

| | | Cost Categories | | | Cost Quantities | | | | | | |
|---|---|---|---|---|---|---|---|---|---|---|---|
| Name | Remarks | Cost Bearer | Life-Cycle | Proj Comp | Qty | UMeas | Unit Cost | Range (±%) | Start Year | End Year | Freq |
| Prestressed I-Beam Type IV (80 ft.) | | agency | ic | super | 20 | ea | $7,000.00 | 0 | 1 | 1 | 1 |
| Overlay concrete to repair deck | | agency | omr | deck | 44 | cy | $1,200.00 | 0 | 40 | 40 | 1 |
| Redirect traffic during deck repair | year 40 | user | omr | deck | 7 | days | | | 40 | 40 | 1 |
| Static load testing of a beam | to insure capacity | agency | ic | new | 1 | ls | 30 000.00 | 0 | 1 | 1 | 1 |

At this point, all initial construction, OM&R, disposal, agency, and user costs have been compiled and inputted into BridgeLCC. The engineer can now analyze the results.

## 6.3    Results

| Step 5: Compute the life-cycle costs of each alternative |

The three ways to view results are with the **Cost Summary** window, the **LCC Summary** and **Timelines** graphs, and the BridgeLCC printed reports (via the **Reports** window).

Figure 41 shows the **Cost Summary** window for the engineer's analysis. Looking at the underlined totals, the life-cycle cost of the conventional concrete bridge is $724,369, while that for the HPC bridge is $675,675. All other things being equal, the HPC bridge is the life-cycle cost-effective design.

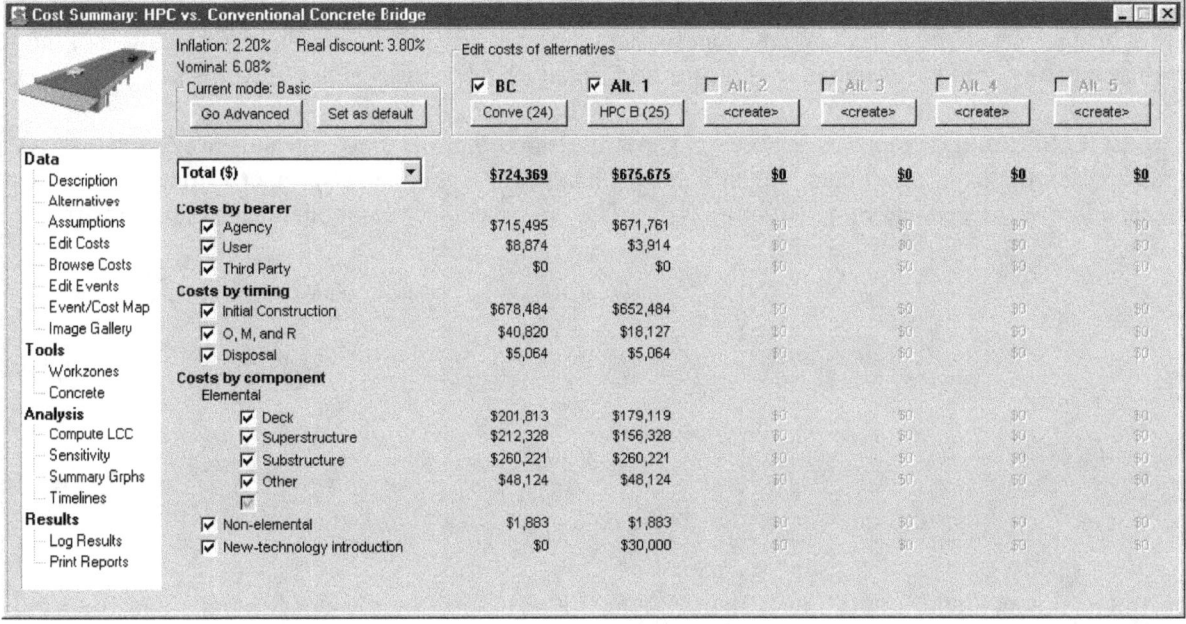

**Figure 41. Example Analysis - Basic: Cost Summary Window**

To investigate where the HPC bridge saves money, we access the **Total $** drop-down list box in the upper-left corner of the panel showing costs and select **Net Savings**. The **Cost Summary** window now shows the net savings of the HPC bridge when compared with the conventional concrete bridge (Figure 42).

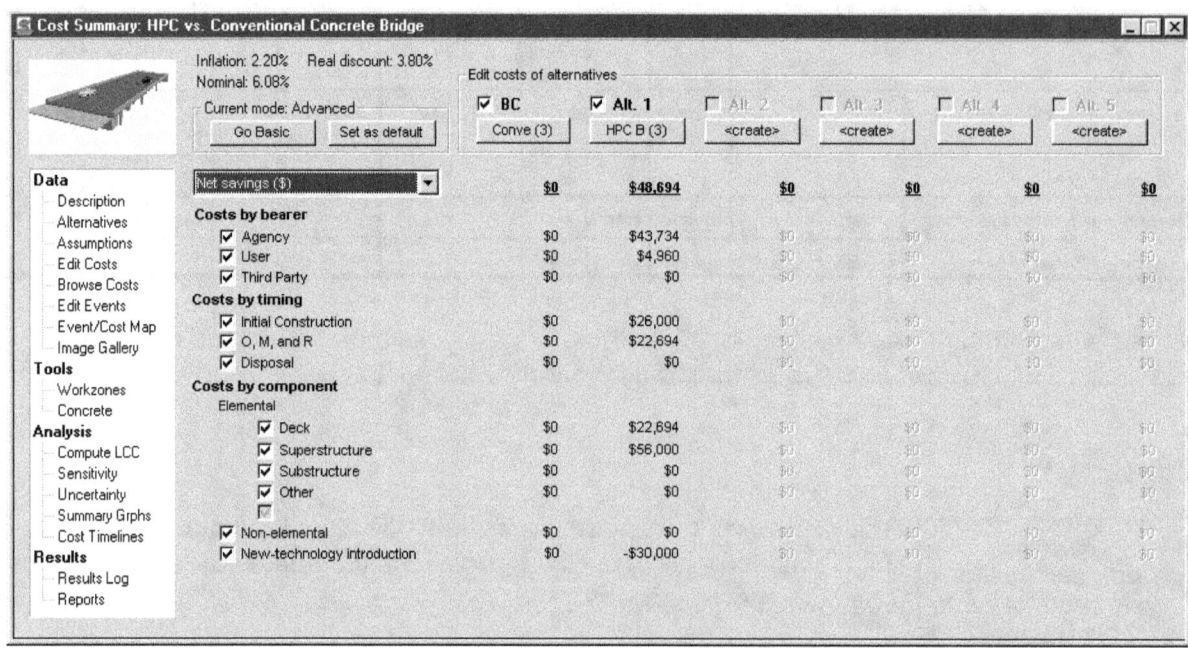

**Figure 42. Example Analysis - Basic: Cost Summary Window Showing Net Savings**

Comparing **Costs by bearer** data, the HPC bridge saves $43,734 in agency costs and $4,960 in user costs (i.e., drivers over the bridge). Comparing **Costs by timing** data, the HPC bridge saves $26,000 in initial construction costs and $22,694 in OM&R costs. Comparing **Costs by component** data, the HPC bridge has savings in deck and superstructure costs, has no savings in substructure and non-elemental costs, and has negative savings (or additional costs) in new-technology costs.

The **LCC Summary** graphs in Figure 43 show the same data in the **Cost Summary** window. As indicated by the front set of bars, the HPC bridge has lower Agency costs, lower Initial Construction costs, and lower Deck and Superstructure costs. The largest project component costs are Substructure costs.

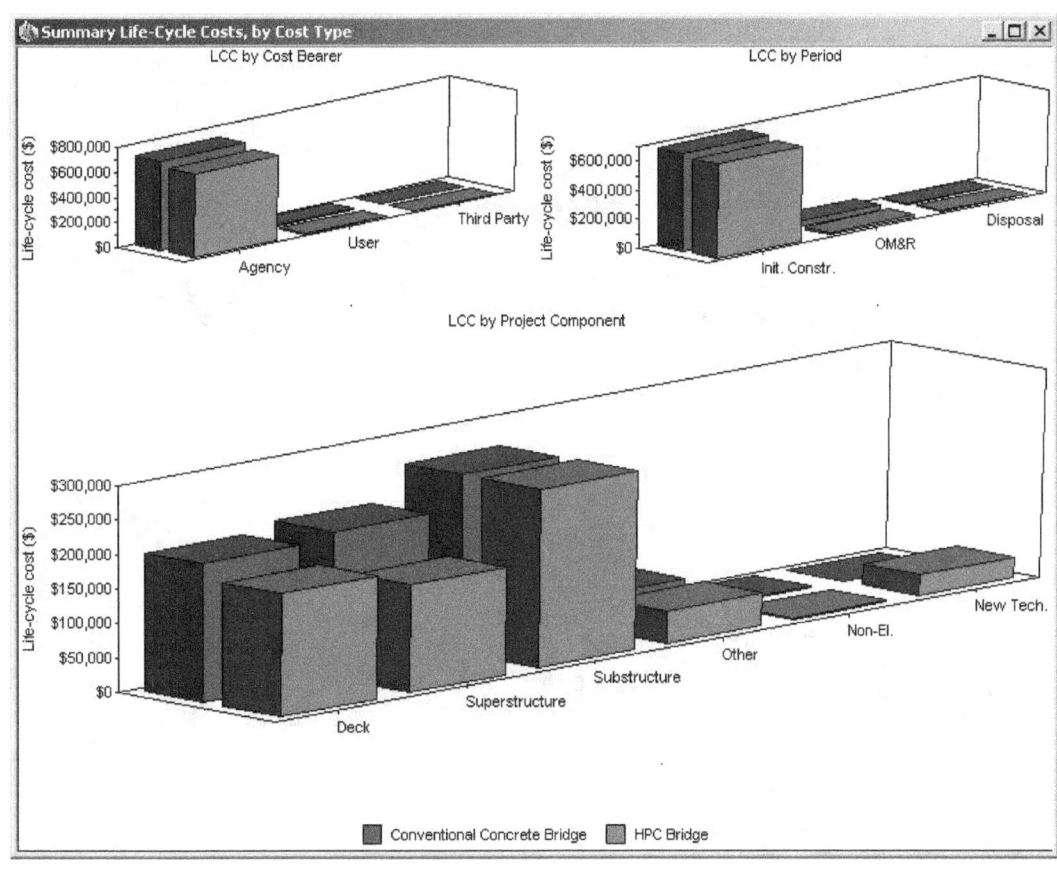

**Figure 43. Example Analysis - Basic: Summary Graph of Life-Cycle Costs, by Alternative**

The cost timelines shown in Figure 44 illustrate the distribution of costs over time. In the upper-left graph, **Yearly Costs in Current-Year Dollars**, we can see that the HPC bridge has high costs in the first year (due to initial construction), the 40th year (due to deck repairs), and the 75th year (due to bridge disposal). The lower-left graph, **Cumulative Costs in Current-Year Dollars**, shows that over the study period the HPC bridge will cost less in cumulative terms; said another way, at no point over the study period will the HPC bridge cost more to-date than the conventional concrete bridge.

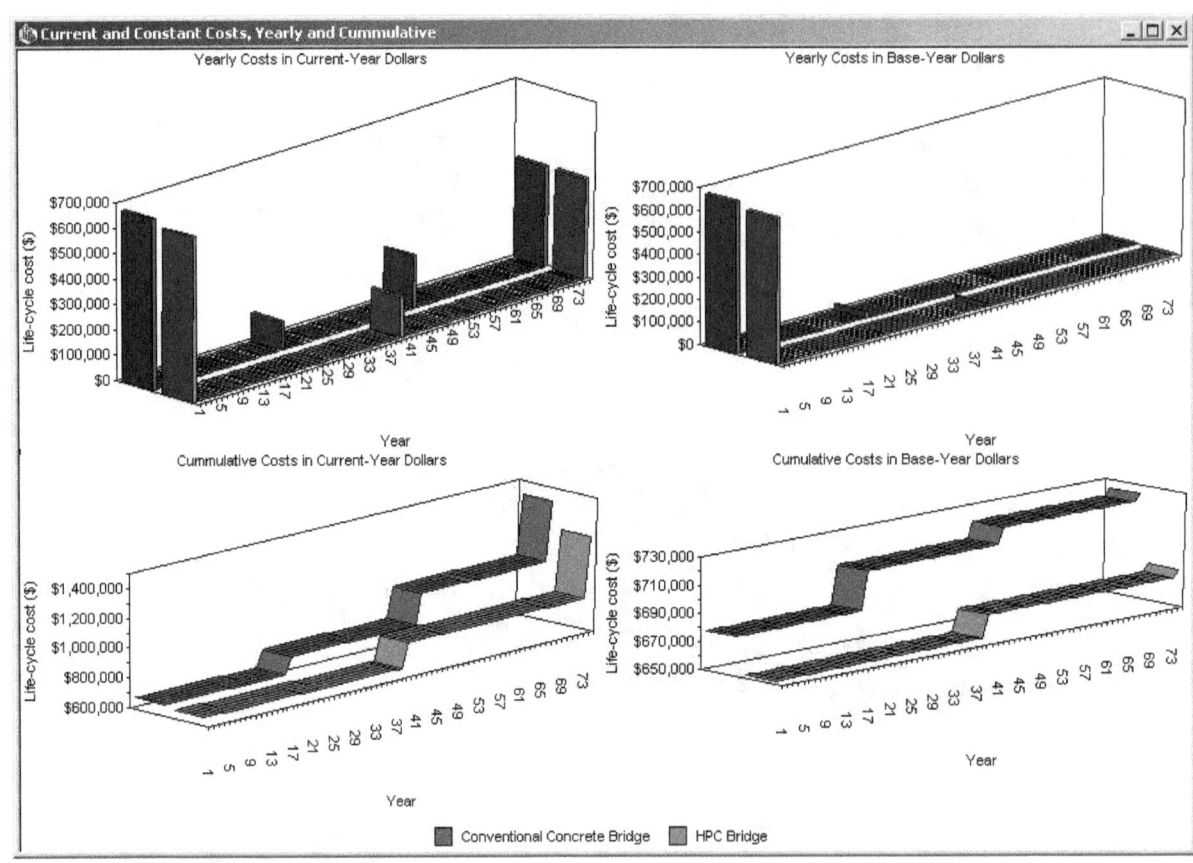

**Figure 44. Example Analysis - Basic: Timelines of Life-Cycle Costs, by Alternative**

The upper-right panel, **Yearly Costs in Base-Year Dollars**, shows the relative importance of constituent cost in the life-cycle cost calculations. Compared with both alternatives' initial construction costs in the first year, the OM&R costs in years 25, 40, and 50 are relatively small – in this particular example and set of interest rates, initial construction costs drive the life-cycle cost competitiveness of each alternative.

**Sensitivity Analysis**

Step 6: Perform sensitivity analysis

Given that the set of best-guess values for each alternative indicates that the HPC bridge is life-cycle cost effective in a deterministic sense, the engineer wants to see whether key underlying parameters affect this result. For example, the HPC bridge may not be cost effective under higher or lower discount rates. To determine the effect of the real discount rate on life-cycle cost, the engineer accesses the **Change in a Single Factor** tab in the **Sensitivity Analysis** window, selects **Discount Rate** from the tree of variables, selects **+/-100%** from the **Variation** drop-down box, and then presses the **Compute** button. The graph in Figure 45 shows the results.

60

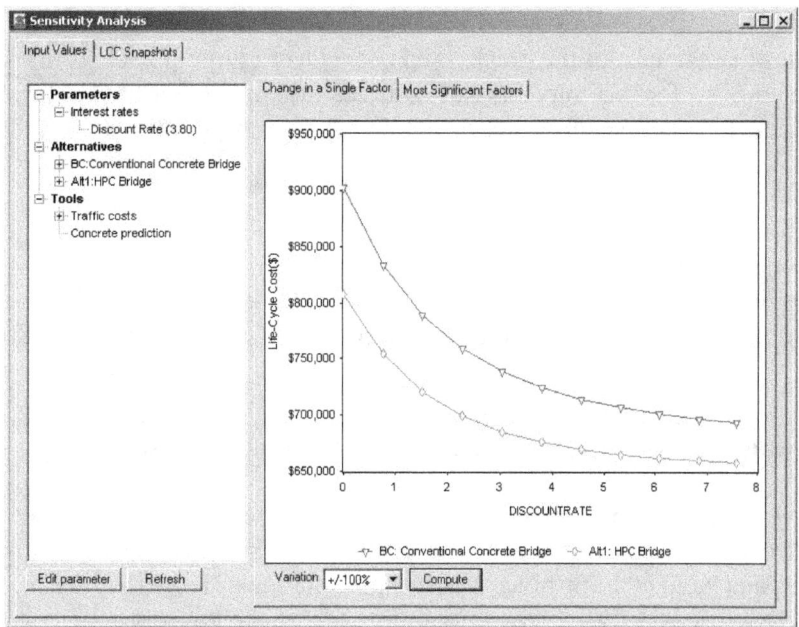

**Figure 45. Example Analysis - Basic: Effect of Real Discount Rate on Life-Cycle Costs**

As indicated by the graph, the life-cycle cost of the HPC bridge is less than that of the conventional concrete bridge when the real discount rate ranges from 0.0% to 7.6%. Said another way, the HPC bridge is life-cycle cost effective regardless of the real discount rate used.

Rather than apply a +/-10% to +/-100% change to every parameter in every alternative, the engineer can get some sense of the relative importance of analysis variables by using the **Most Significant Factors** tab in the **Sensitivity Analysis** window to test the effect of 10% changes in variables on life-cycle costs. Figure 46 shows one of the graphs that can be displayed after the **Most Significant Factors** tab computes the relative importance of analysis variables.

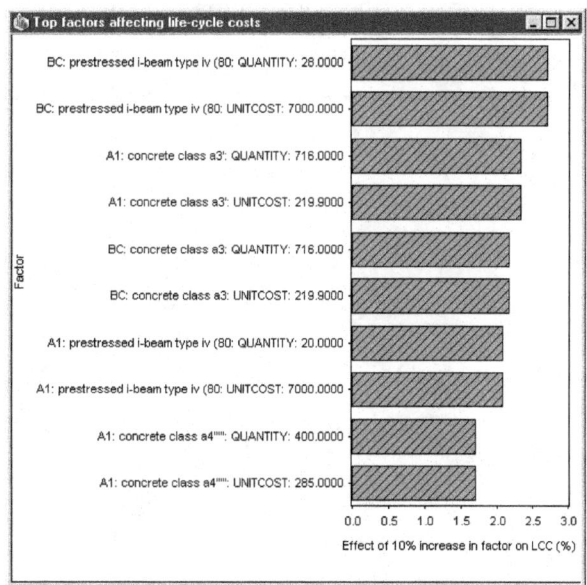

**Figure 46. Example Analysis – Basic: Graph of Top Factors Affecting Life-Cycle Costs**

61

The figure indicates that the cost of the prestressed beams has a high impact on the life-cycle costs of both alternatives, more so for the conventional-concrete bridge since it has 28 beams instead of 20. The unit costs of concrete for the substructure and deck have strong influence on the life-cycle costs of both alternatives.

At least two useful results come out of the **Most Significant Factors** tab. First, the engineer knows that the factors toward the bottom of this generated list have little effect on life-cycle cost and thus do not warrant further analysis. Second, if the engineer is looking to reduce the life-cycle costs of either alternative design, he or she can look to the top factors on this list and seek ways of reducing their costs.

---

**Optional Advanced-Mode Subject: Uncertainty and Risk Analysis**

As an example of the use of Monte Carlo simulations, we can optionally assume that the engineer has relative uncertainty about the HPC bridge's unit costs, and thinks that this uncertainty may prevent it from always being the life-cycle cost-effective bridge material. For example, suppose that there is a probability that the HPC-bridge unit costs of initial construction, OM&R, and disposal costs are all much higher than the base-case bridge costs; in this case there is a small probability that the HPC bridge will not be life-cycle cost effective.

The engineer tests a specific example. Suppose he or she believes that the costs for the base case bridge can vary as much as 5% from the best-guess values, while the HPC bridge unit costs can vary as much as 10% from their best-guess values. For each of the base case costs, the engineer applies a 5% uncertainty to each unit cost in the base case and applies a 10% uncertainty to the unit costs in the HPC bridge.

To conduct uncertainty and risk analysis of this relative uncertainty, the engineer accesses the **Uncertainty and Risk** window, sets a number of samples for the Monte Carlo simulation in the **Run Simulation** tab, and presses the **Run** button. Upon completion, access the **View Results** tab (Figure A).

---

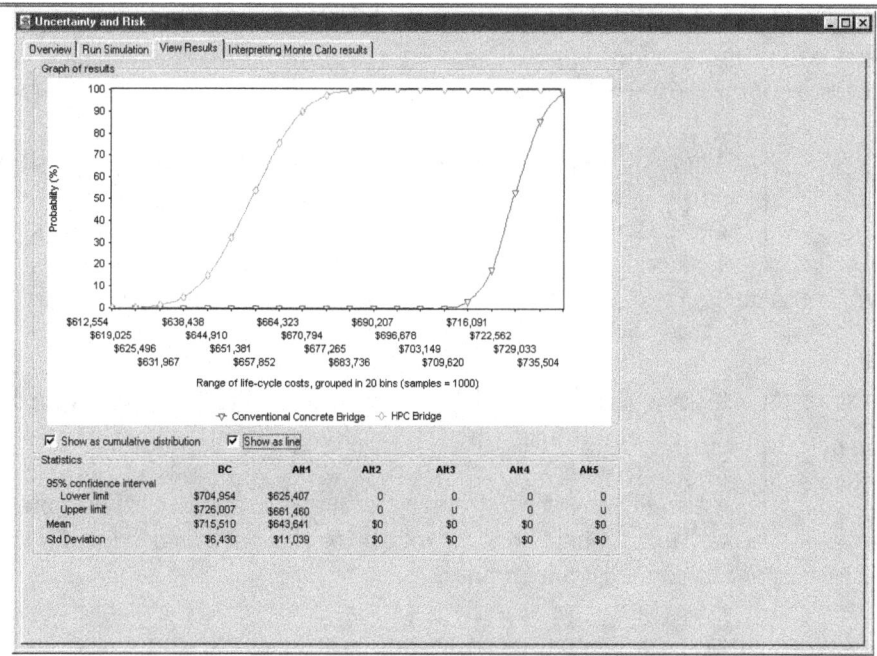

**Figure A. Example Analysis - Basic: Monte Carlo Simulation Results**

The graph shows the range of possible life-cycle costs for each alternative. Moving from left to right across the graph, we see that, for each probability level, Alt #1 (the HPC bridge) has a lower expected life-cycle cost than the Base Case (conventional-concrete bridge). For example, 10% of the time (ten times out of 100) the HPC-bridge life-cycle cost is less than or equal to $640,000, while 10% of the time the base-case bridge life-cycle cost is less than or equal to $720,000. Said another way, the cost of the HPC bridge will be at least as low as $640,000, while the cost of the base-case bridge will be at least as low as (the higher) $720,000. The HPC bridge has a lower possible life-cycle cost.

Since for every probability level the HPC bridge has a lower expected life-cycle cost than the base-case bridge, the HPC bridge is cost effective in a probabilistic sense. In a statistical sense, the HPC bridge cost distribution *strictly dominates* the conventional-concrete bridge cost distribution.

If one alternative's distribution does not strictly dominate all other alternatives distributions, none of the material alternatives would be cost effective in a probabilistic sense. Some additional procedures for including risk attitude in project evaluation would be required to establish the preferred bridge.[6]

---

| Step 7: Compare the alternatives' life-cycle costs |
| --- |

The engineer now compares the alternatives based on their best-guess life-cycle costs in the **Cost Summary** window and on the three sensitivity tests. The HPC bridge is cost-effective based on best-guess values: $675,675

---

[6] For more discussion about risk attitudes and uncertainty, see Rosalie T. Ruegg and Harold E. Marshall, *Building Economics: Theory and Practice* (New York, New York: Chapman and Hall, August 1990).

versus $724,369. The sensitivity analysis allows the engineer to conclude that the HPC bridge is life-cycle cost effective (1) for a wide range of interest rates and (2) when the base case and HPC bridges' costs vary by 5% and 10%.

| Step 8: Consider other project effects | Factors other than cost can affect an engineer's design about what material to use. These non-cost factors could include architectural considerations, material restrictions, or politics. An engineer can use additional procedures such as the multi-attribute decision analysis (MADA) to weigh cost and |

non-cost factors simultaneously.[7] In this example analysis, only cost affects the final material decision.

| Step 9: Choose the life-cycle cost-effective alternative | Given that all cost and non-cost factors have been considered, the engineer can conclude that the HPC bridge is life-cycle cost-effective when compared with his base-case, conventional-concrete bridge. Its life-cycle cost is lower than the other alternatives, and sensitivity analysis indicates that this conclusion is robust to the selected changes in underlying |

parameters and assumptions about cost uncertainty.

This completes the analysis. To summarize, the engineer analyzed the life-cycle costs of two alternative designs: (a) a base-case bridge made from conventional-strength concrete and (b) an alternative bridge made from new, high-performance concrete. Both designs met the engineer's requirements, including design codes and service life. The engineer compiled the costs of building, maintaining, and eventually disposing of each bridge.

Using government-set inflation and real discount rates, the engineer then computed the life-cycle costs of each alternative, and found the HPC bridge to be the life-cycle cost-effective alternative. Next, the engineer found that the HPC bridge was life-cycle cost effective for real discount rates that ranged from 0.0% to 7.6%, that is, the real discount rate had no effect on the life-cycle cost-effective choice. Finally, the engineer computed the top factors affecting the life-cycle costs and found that deviations in the costs of concrete for the alternatives' beams, decks, and substructures costs had the strongest impacts on changing life-cycle costs.

Under the particular set of sensitivity values, HPC is the cost-effective bridge material. HPC allows the designer to use fewer beams and to have a smaller repair schedule over the life of the structure. This saves the agency construction and repair costs and saves drivers on the highway both time and cost.

---

[7] For a detailed description of MADA techniques, see Gregory Norris and Harold E. Marshall, *Multi-attribute Decision Analysis Method for Evaluating Buildings and Systems*, 1995.

# 7. Example Analysis – Advanced: Terrorism Risk Management

This chapter illustrates some of the advanced features in BridgeLCC, including the use of *events* that can have probabilities of occurring. Compared with the example analysis in the previous section, less detail is given about the alternatives' designs and site conditions and more about the expected outcomes and associated costs of man-made and natural attacks to civil infrastructure. This analysis can be accessed in the file "Terrorism Risk Management.lcc," located in your BridgeLCC directory.

In this example an engineer is assessing the life-cycle cost effectiveness of three alternative designs:

1. A base-case design that has no designed resistance to terrorist or seismic attack,
2. An alternative design specifically designed to reduce damage from terrorist attack, and
3. An alternative design specifically designed to reduce damage from terrorist and seismic hazards.

In each case, there is a probability every year in the study period that one of three things may occur: (1) a terrorist will make an unsuccessful attack on the bridge, resulting in minor damage to the structure and minor delays and costs for drivers/users of the bridge; (2) a terrorist will make a successful attack on the bridge, resulting in major damage to the bridge and major delays and costs for drivers/users of the bridge; and (3) no terrorist attack will occur. In each case the engineer makes estimates of the costs associated with each outcome, for each alternative design.

To estimate the life-cycle costs of each alternative, BridgeLCC *events* are constructed and assigned probabilities of occurring. The engineer then assigns costs to each possible outcome. Since some events are probabilistic, BridgeLCC computes *expected* life-cycle costs.

To estimate the *range of possible* life-cycle costs of each alternative that results from such probabilities, the engineer performs Monte Carlo simulations, where specific outcomes are sampled from the probabilities, resulting in a wide range of possible outcomes. In some cases no attacks occur, whereas in some years there are multiple attacks. Still, the engineer can determine whether any one of the three alternatives is life-cycle cost effective in a probabilistic sense.

As with the previous chapter, this analysis is broken into three sections: (1) an overview of the analysis process and objectives, (2) a description of the project data, and (3) an interpretation of the life-cycle cost results. The analysis is again viewed from the perspective of a design engineer who must assess which of his or her alternative designs is the life-cycle cost-effective choice.

## 7.1    Overview

| Step 1: Define the project objective and minimum performance requirements | The project objective is to build, maintain, and eventually dispose of a new highway bridge. The structure must satisfy all pertinent design codes and last at least 75 years. |
| --- | --- |

<table>
<tr><td>

**Step 2:**
**Identify the**
**alternatives for**
**achieving the**
**objective.**

</td><td>

The engineer is considering three alternative bridge designs. The first, or base-case, design has no particular improvements to reduce damage from large seismic loading or man-made attack. The second alternative design specifies improvements that will reduce damage from large seismic loads, but at some additional costs. The third and final alternative specifies (1) design improvements that will reduce damage from large seismic loads and (2)

</td></tr>
</table>

enhanced security that should reduce the likelihood of a successful man-made attack. The engineer must determine which alternative is the life-cycle cost-effective design.

## 7.2    Data

### Project Parameters

<table>
<tr><td>

**Step 3:**
**Establish the basic**
**assumptions for**
**the analysis.**

</td><td>

The three sets of conditions that are common to the three alternatives are: terrorist conditions, traffic conditions, and economic conditions. While man-made attack conditions might not be considered to be the same for all three alternatives – terrorists or other attackers might be less likely to attack a structure that has enhanced security and design; for this example we are assuming the probabilities are the same.

</td></tr>
</table>

The engineer divides the range of possible attack events into groups that correspond to estimated categories of bridge damage, such as no damage, minor damage, major damage, or collapse. Table 8 lists the categories of attack, their probabilities of occurring, and related levels of damage. Each year, the bridge will experience one of these outcomes (typically, no attack).

**Table 8. Terrorist Attack Frequencies and Damage Levels, by Alternative**

| Outcome | Probability of occurring (%) | Base case: no enhanced seismic or security design | Alternative 1: enhanced seismic design | Alternative 2: enhanced seismic and security design |
|---|---|---|---|---|
| No attack | 99.94% | No damage | No damage | No damage |
| Unsuccessful attack | 0.05% | Major damage | Minor damage | Minimal damage |
| Successful attack | 0.01% | Very major damage | Major damage | Some damage |

Next, the engineer tallies the traffic conditions that the bridge will experience over the study period. When the bridge is constructed, undergoing maintenance-related repair, or undergoing repair from a seismic or terrorist event, the drivers who use the bridge will experience driver delay costs, vehicle operating costs, and increased accident costs. (The formulas that BridgeLCC uses to compute these costs are listed in Appendix B.)

Table 9 lists the common workzone conditions that will be needed for constructing or repairing one or more of the alternative bridges. (See the text below Table 2 on page 50 for definitions of these traffic parameters.) Two of the parameters, length of affected workzone and roadwork driving speed, vary according to the amount of repair or damage to the alternative bridge.

## Table 9. Workzone Parameters

| Item | Value |
|---|---|
| Average Daily Traffic (ADT) | 2003: 40,000 |
|  | 2078: 60,000 |
| Length of affected roadway (miles) | Varies |
| Normal driving speed | 55 mph |
| Roadwork driving speed | Varies |
| Normal accident rate (per million veh miles) | 1.9 |
| Roadwork accident rate (per million veh miles) | 2.2 |
| Hourly driving cost (base year $) | $5.00 |
| Hourly vehicle operating cost (base year $) | $10.00 |
| Cost per accident ($) | $100,000 |

These data are inputted into the **Workzones** tab of the **Project Assumptions** window.

The final set of parameters contains the inflation rate and real discount rate. The inflation rate of 2.2% is taken from the Bureau of Economic Analysis. The real discount rate used is the same as that listed in Table 4 of the previous example analysis. For clarity, the rates are listed again in Table 10.

## Table 10. Real Discount Rate, by Number of Years

| Number of Years | Discount Rate (%) |
|---|---|
| 7 years | 3.0 % |
| 30 years | 3.8 % |

This rate and the inflation rate are inputted into the **Economic** tab in the **Project Assumptions** window.

## Events

Each possible attack outcome has a probability attached to it as well as a set of costs, two conditions that make necessary the use of BridgeLCC *events*. Access the **Edit Events** window to input the events. First, an independent "Annual attack-related event" is inputted to define that one of the three possible mutually exclusive outcomes will happen each year. Next, three events that are dependent on the "Annual" event are created: (1) a "No attack" event with probability 99.94%, (2) an "Unsuccessful attack" event with probability 0.05%, and (3) a "Successful attack" event with probability 0.01%. Figure 47 shows the inputted events.

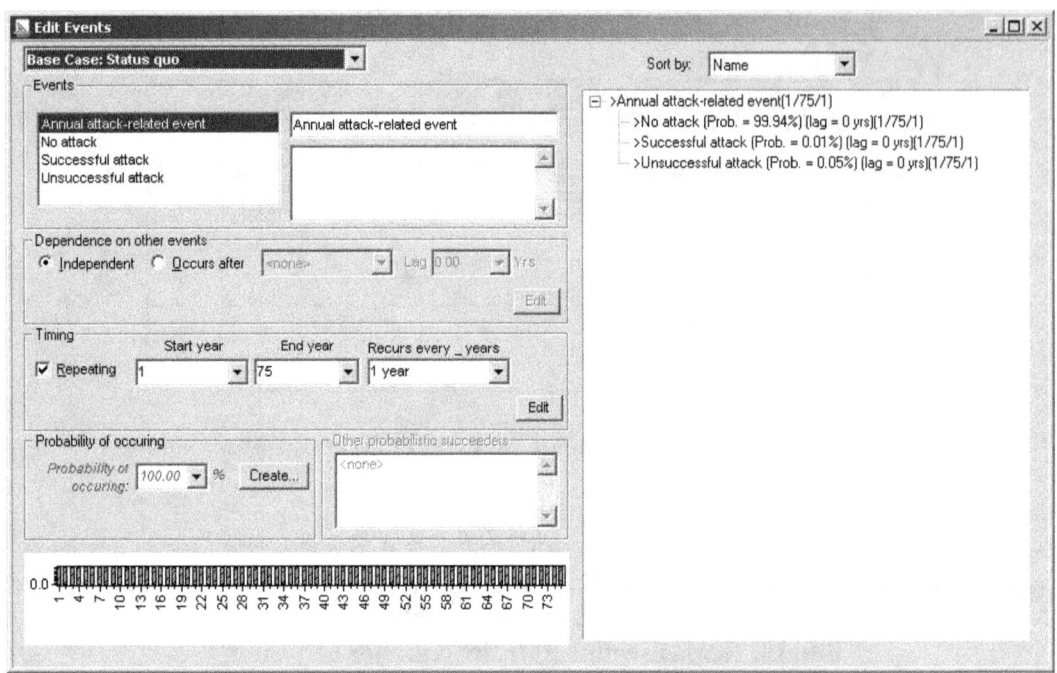

**Figure 47. Example Analysis - Advanced: Edit Events Window**

| Step 4:<br>Identify, estimate,<br>and determine the<br>costs | Next, the engineer itemizes the set of costs of bridge construction, repair, and disposal over the study period. Some of these costs occur in specific years, such as during initial construction in the base year, while others only occur if a specific event occurs. |

The engineer organizes costs according to who bears the costs (*agency, user, and third-party costs*), when the costs occur (*initial construction; operation, maintenance, and repair [OM&R], and disposal*), and to what part of the structure the costs are tied (e.g., *deck*). Table 11 lists the agency costs for each alternative.

**Table 11. Example Analysis - Advanced: Agency Costs, by Alternative**

| Project: Terrorism risk management | | | | Date: October 27, 2003 | | | Page: 1 of 1 | | |
|---|---|---|---|---|---|---|---|---|---|
| Remarks: agency costs for each alternative | | | | | | | | | |
| | Cost Categories | | | Cost Quantities | | | | | |
| Name | Cost bearer | Life cycle | Proj. Comp. | Qtty | Unit. Meas. | Unit cost | Start Year | End Year | Freq. |
| **Base case** | | | | | | | | | |
| Initial construction | agency | i.c. | non-el | 1 | LS | $1,000,000 | 1 | 1 | 1 |
| Repair: unsucc. attack | agency | omr | non-el | 1 | LS | $100,000 | Event: unsuccessful attack | | |
| Repair: succ. attack | agency | omr | non-el | 1 | LS | $1,000,000 | Event: successful attack | | |
| | | | | | | | | | |
| **Alternative 1** | | | | | | | | | |
| Initial construction | agency | i.c. | non-el | 1 | LS | $1,150,000 | 1 | 1 | 1 |
| Repair: unsucc. attack | agency | omr | non-el | 1 | LS | $75,000 | Event: unsuccessful attack | | |
| Repair: succ. attack | agency | omr | non-el | 1 | LS | $150,000 | Event: successful attack | | |
| | | | | | | | | | |
| **Alternative 2** | | | | | | | | | |
| Initial construction | agency | i.c. | non-el | 1 | LS | $1,200,000 | 1 | 1 | 1 |
| Repair: unsucc. attack | agency | omr | non-el | 1 | LS | $25,000 | Event: unsuccessful attack | | |
| Repair: succ. attack | agency | omr | non-el | 1 | LS | $25,000 | Event: successful attack | | |

Once tabulated, these costs are inputted into BridgeLCC using either the **Edit Costs** window or the **Browse All Costs** window, the latter of which is shown as an example in Figure 48. Note in the last **Event** column of the window that each cost is tied to its respective event.

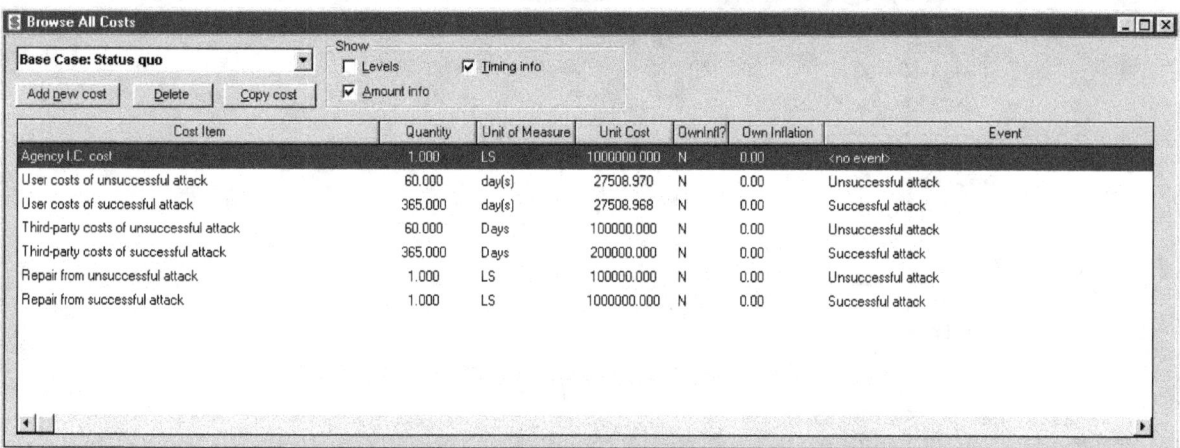

**Figure 48. Example Analysis - Advanced: Browse All Costs Window**

To view a complete listing of all events in an alternative and the costs tied to each, access the **Event/Cost Map** window, shown in Figure 49. Use this window to insure that all events have been input correctly, and that all costs have been input and are tied to the appropriate event (or to no event).

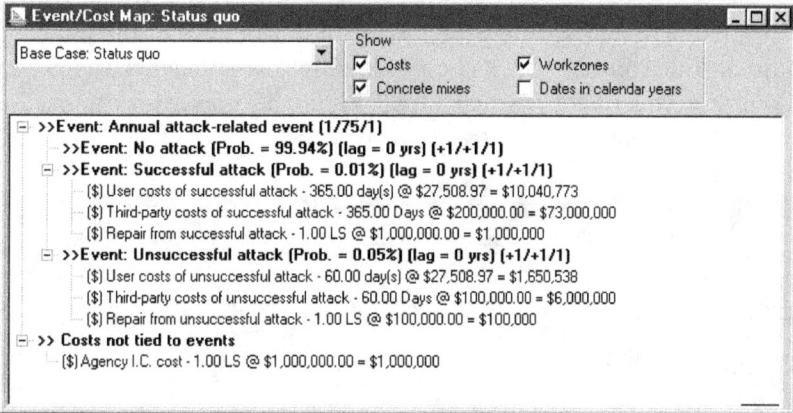

**Figure 49. Example Analysis - Advanced: Event/Cost Map**

At this point, the input of data has been completed. To insure that all data have been input correctly, print the data-related reports, either by selecting **File/Print...** from the menu or by accessing the **Cost Summary** window and selecting **Reports** underneath the **Results** heading in the left panel.

## 7.3    Results

Step 5:
Compute the life-
cycle costs of each
alternative

Once the events, workzones, economic parameters, and costs have been correctly input, the life-cycle costs of each alternative can be viewed in the **Cost Summary** window (Figure 50). Since the data include attack-related events that have some less-than-100% probability, the values listed in the window are the *expected-value life-cycle costs*; some of the calculations are the costs of the event (say, a successful attack) multiplied by the probability of the event occurring (in our case, 0.01%).

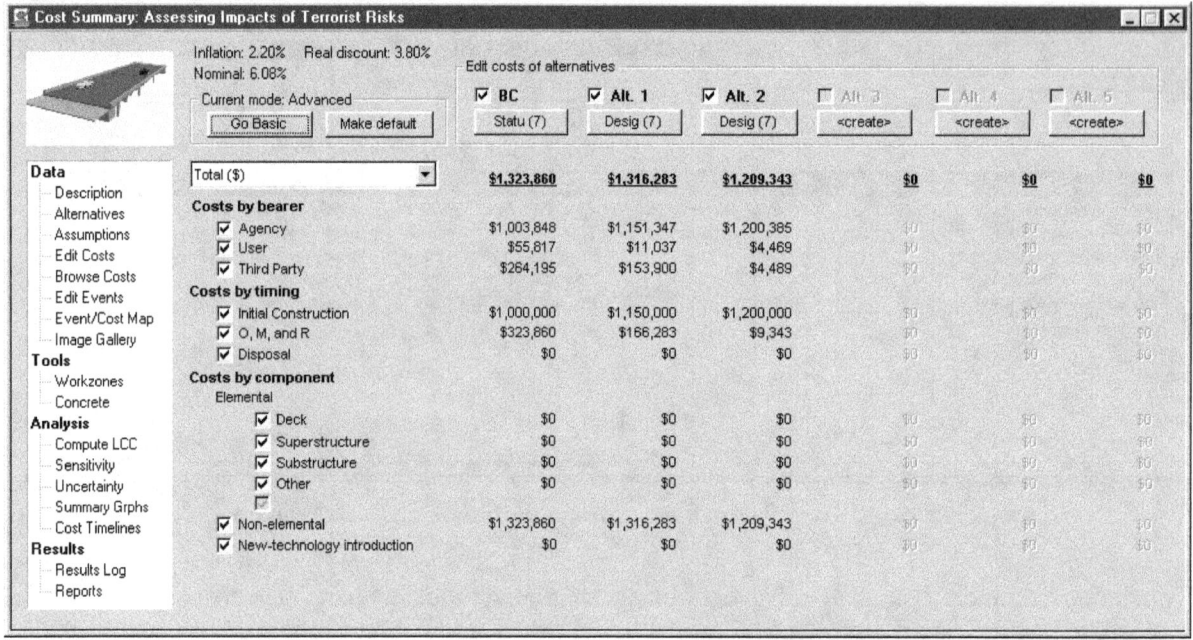

**Figure 50. Example Analysis - Advanced: Cost Summary Window, Total $**

As indicated in Figure 50, Alternative 2 has the lowest life-cycle cost of the three alternatives. While it has higher initial construction costs, it has considerably lower repair costs. Figure 51 makes the same comparison, but in terms of $ per square foot; Figure 52 compares the net savings per square foot of Alternatives 1 and 2 when compared with the base case.

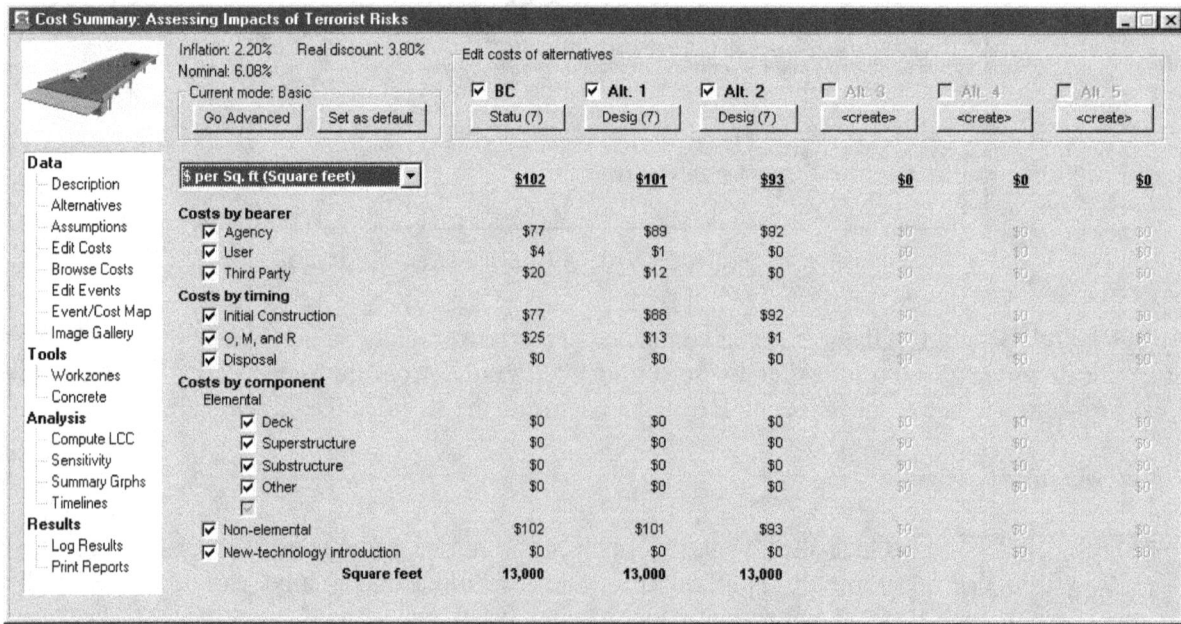

**Figure 51. Example Analysis - Advanced: Cost Summary Window, $ per Sq. Feet (Area of Deck)**

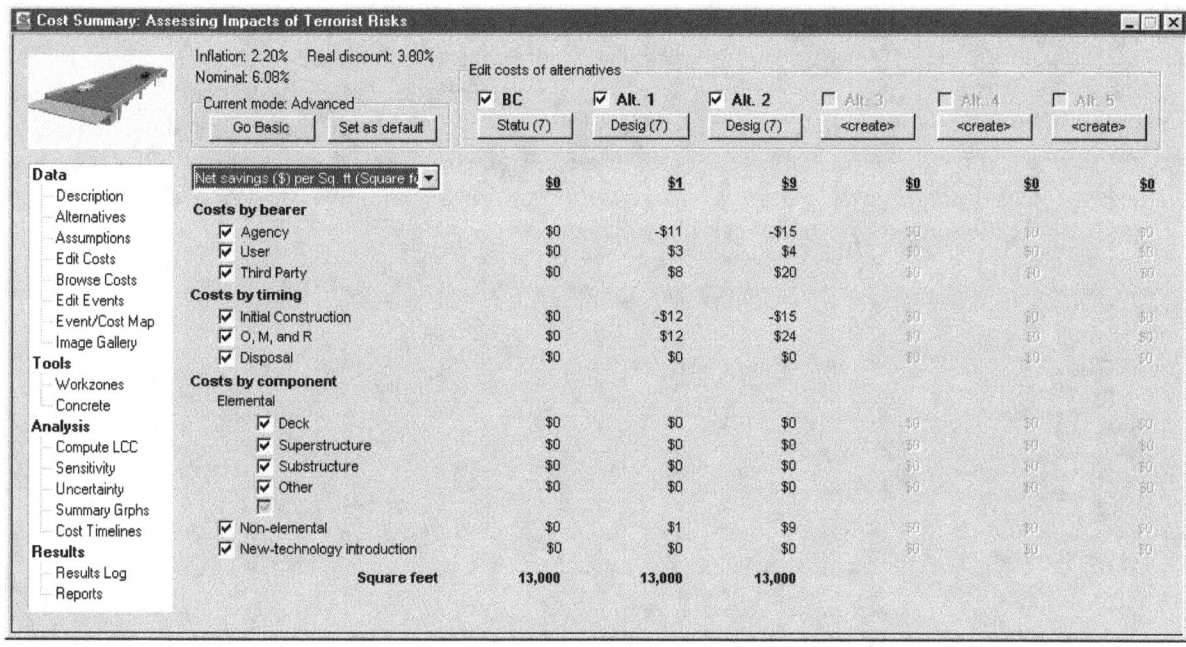

**Figure 52. Example Analysis - Advanced: Cost Summary Window, Net Savings ($) per Sq. Feet**

Figure 53 illustrates the timing of costs over the study period. All three alternatives have high annual costs in the first year, reflecting initial construction, and the base case alternative has relatively high (expected) annual costs over the study period; in total, these expected annual terrorism costs prevent the alternative from being life-cycle cost effective.

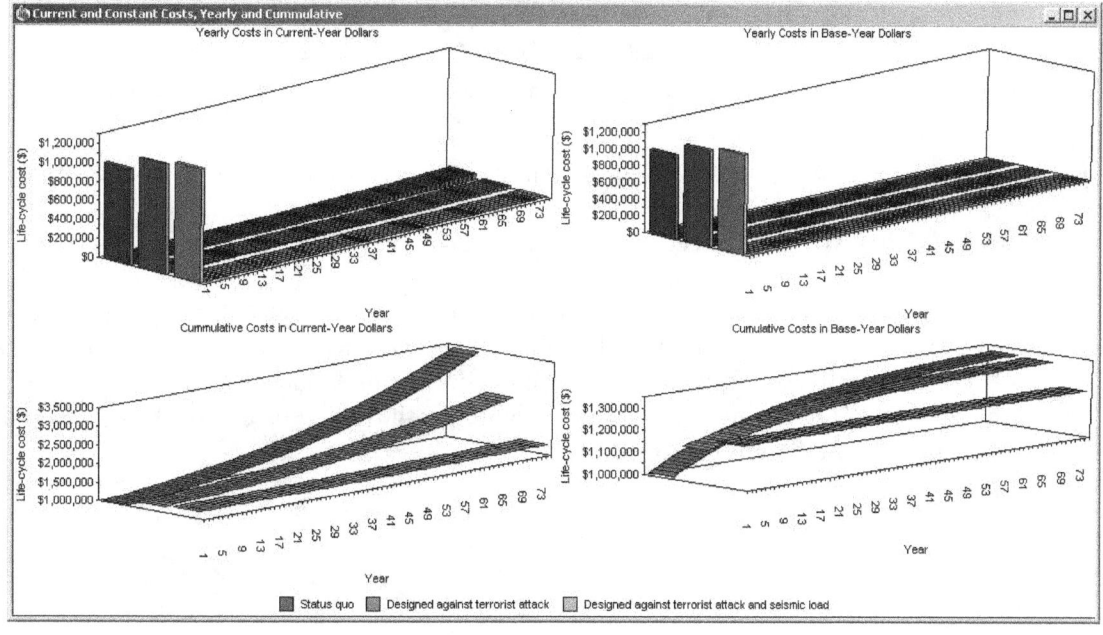

**Figure 53. Example Analysis - Advanced: Timelines of Costs**

## Sensitivity Analysis

| Step 6: Perform sensitivity analysis |
| --- |

To assess the importance of the real discount rate on the determination of the life-cycle cost-effective alternative, the engineer accesses the **Sensitivity Analysis** window, selects **Discount Rate** from the left tree of parameters, selects a variation from the **Variation** drop-down list box, and presses the **Compute** button. Figure 54 shows the resulting graph of life-cycle costs for each alternative over the range of possible real discount rates.

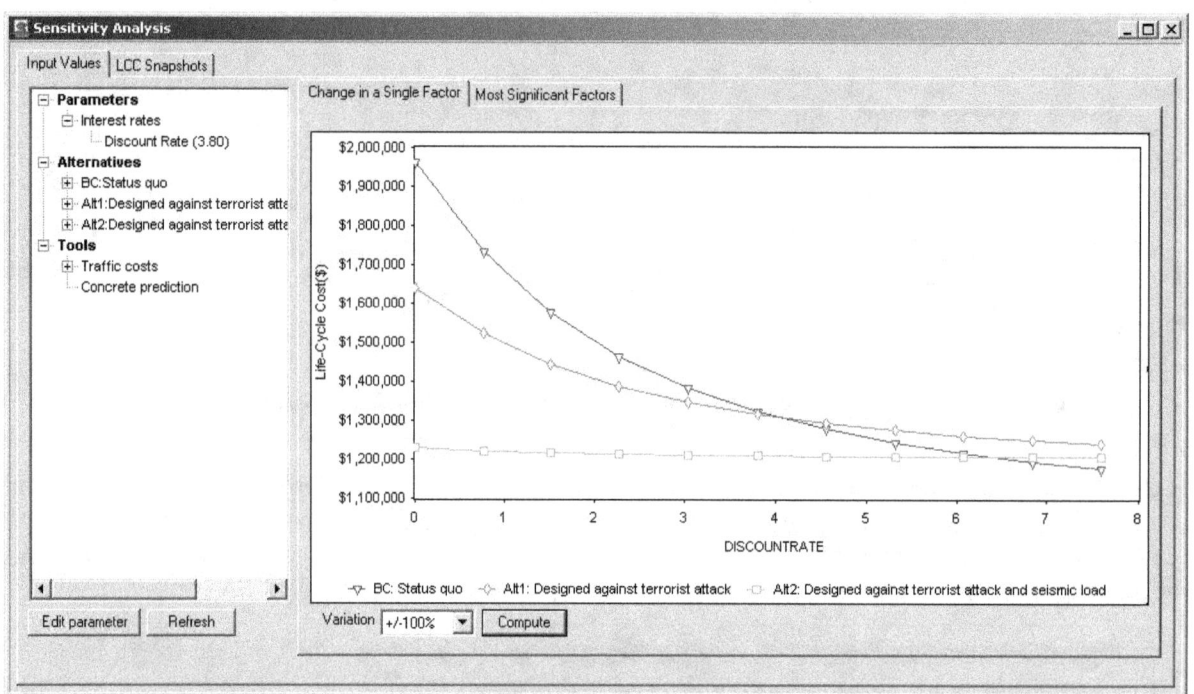

**Figure 54. Example Analysis - Advanced: Change in Single Factor, Real Discount Rate**

The figure illustrates that Alternative 2 is life-cycle cost effective over the 0% to 6.5% range of interest rates; said another way, the real discount rate used in this range does not affect the determination of the life-cycle cost-effective bridge design.

Finally, the bridge engineer has relative uncertainty about the unit prices for the repairs due to terrorist attacks. To model this uncertainty, he assigns a 25% uncertainty to the terrorist-related repairs in all three alternatives, and then runs a Monte Carlo simulation of the range of life-cycle costs that can result from this range of unit prices. Figure 55 shows the results of the simulation, in the **View results** tab in the **Uncertainty and Risk** window.

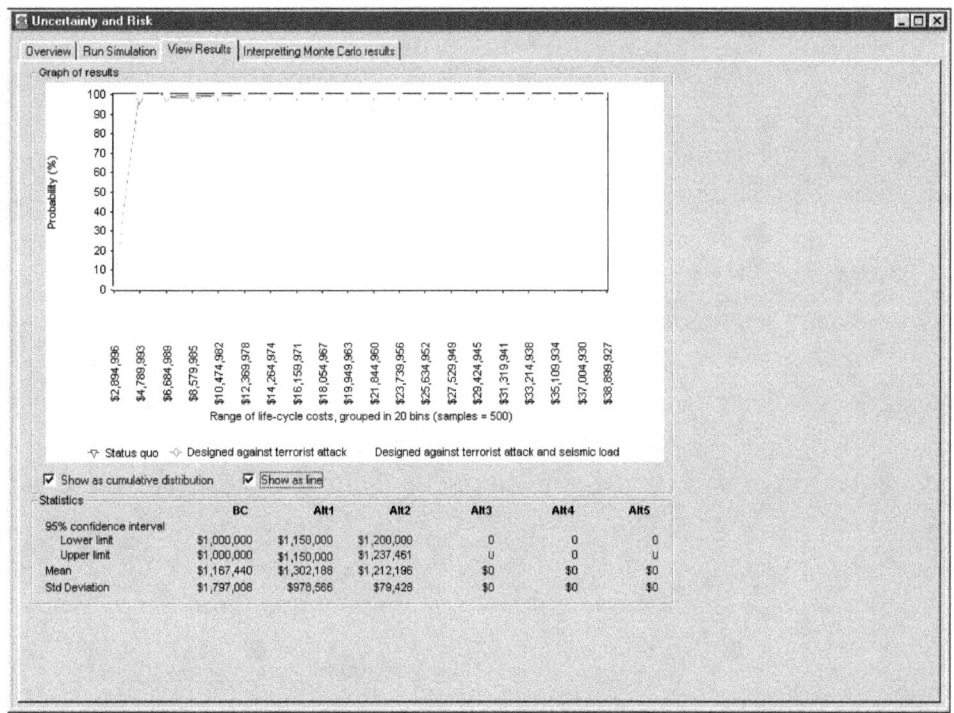

**Figure 55. Example Analysis - Advanced: Monte Carlo Simulation Results**

The graph shows the cumulative probability of each alternative over the range of life-cycle cost outcomes for all alternatives. Since the cumulative probability of Alternative 2 is always above the other two alternatives' lines, Alternative 2 statistically *strictly dominates* the other two and therefore is life-cycle cost effective in a probabilistic sense.

| | | |
|---|---|---|
| **Step 7: Compare the alternatives' life-cycle costs** | Since Alternative 2 is both life-cycle cost effective in a deterministic sense (it has the lowest expected life-cycle cost of the three, as shown in the **Cost Summary** window) and is life-cycle cost effective in a probabilistic sense (it strictly dominates the other two distributions in the **View results** tab), it is the life-cycle cost-effective alternative. (If both conditions do not hold for | |

one of the three alternatives, additional techniques are needed.)

| | |
|---|---|
| **Step 8: Consider other project effects** | As discussed in the previous example, factors other than cost can affect an engineer's design about what material to use. These non-cost factors could include architectural considerations, material restrictions, or politics. An engineer can use additional procedures such as the multi-attribute decision |

analysis to weigh cost and non-cost factors simultaneously. In this example analysis, only cost affects the final material decision.

| | |
|---|---|
| **Step 9: Choose the life-cycle cost-effective alternative** | Given that all cost and non-cost factors have been considered, the engineer can conclude that Alternative 2 is life-cycle cost effective; its life-cycle cost is lower than the other alternatives, and sensitivity analysis indicates that this conclusion is robust to the selected changes in underlying parameters and assumptions about cost uncertainty. |

# 8. Additional Example Analyses

In addition to the two examples described in Chapters 6 and 7, BridgeLCC includes some additional, smaller analyses that help explain details about some of the features in the software.

**Comparing Workzones** ("Comparing Workzones.lcc") – this example compares the user costs associated with two alternative workzones during the repair of a 4-lane one-way bridge: (1) a workzone that closes one lane of traffic, causing traffic to move over only the three remaining lanes, and (2) a workzone that closes two lanes.

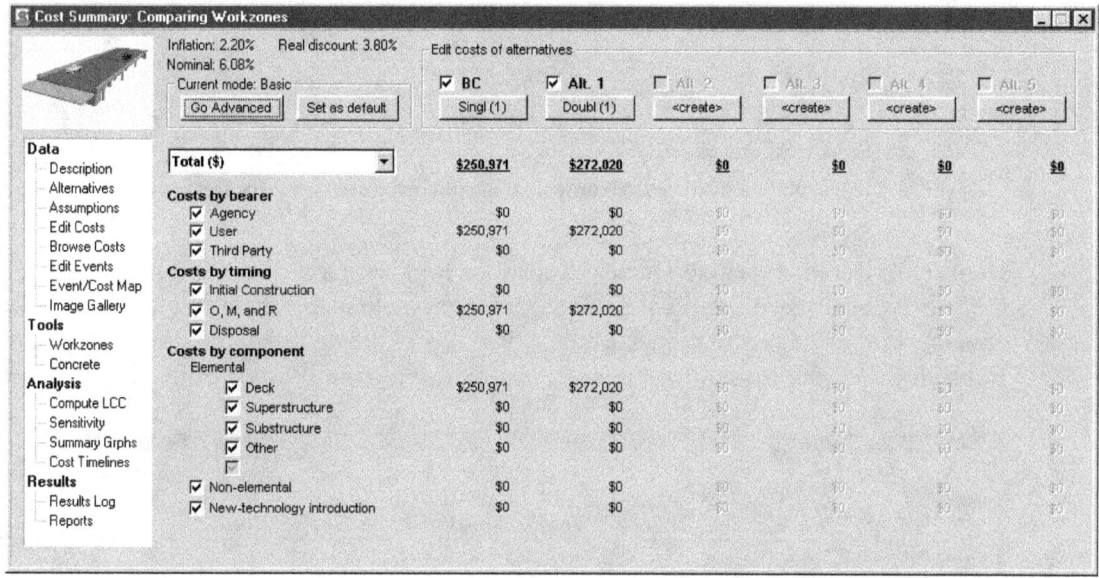

**Figure 56. Additional Example: Comparing Workzones**

**Comparing Concrete Mix Designs** ("Concrete Mix Designs.lcc") – this example uses the BridgeLCC Concrete Service Life Prediction Tool to compare two alternative mix designs over the life-cycle of a bridge that is exposed to high levels of road salts: (1) a base case, conventional mix design typically used in highway bridges and (2) a high-performance concrete (HPC) mix design which has significant resistance to road salts. The HPC design significantly reduces the time between repairs to the bridge deck, thereby reducing repair costs and overall life-cycle costs.

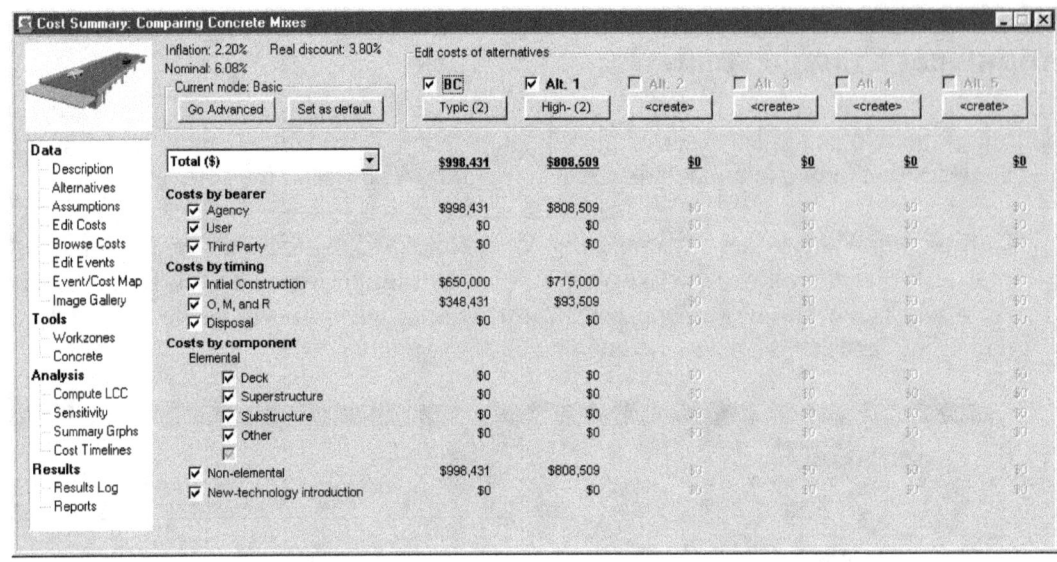

**Figure 57. Additional Example: Comparing Concrete Mix Designs**

**Repair or Replace Bridge Deck** ("Repair or Replace Deck.lcc") – this example compares the life-cycle costs of a typical bridge decision: whether to repair or replace the deck of an existing bridge. Construction costs and user costs are ignored for the sake of exposition. Replacing the deck is found to be the cost-effective alternative over a 75-year study period.

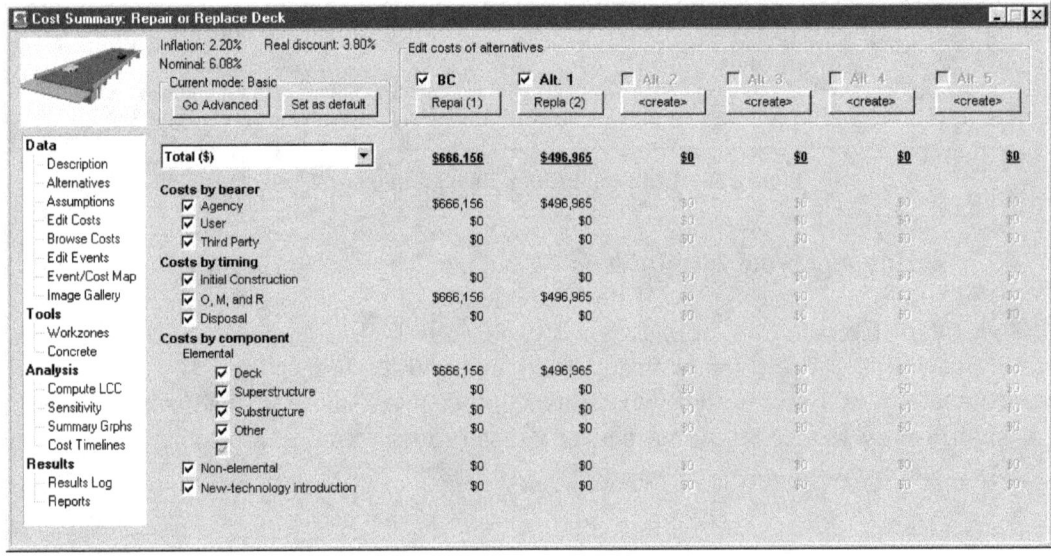

**Figure 58. Additional Example: Repair or Replace Bridge Deck**

# Appendix A.  Life-Cycle Costing Methodology and Cost Classification Scheme

BridgeLCC uses a life-cycle costing methodology based on the ASTM practice E 917 for measuring the life-cycle costs of buildings and building systems and a cost classification scheme developed by Ehlen and Marshall (1996). The classification scheme in particular allows the user to capture all project-related costs and compare alternatives' life-cycle costs in useful ways. BridgeLCC's online help gives definitions of each cost type.

This appendix gives an abridged description of the methodology and classification used in BridgeLCC.

## A.1    The Life-Cycle Costing Methodology

### Steps in Life-Cycle Cost Analysis

The recommended steps for calculating the life-cycle cost of a new-technology material vis-à-vis a conventional material are as follows:

1. **Define the project objective and minimum performance requirements.** The performance requirements of a project should be expressed in terms that do not preclude the use of a new-technology material.

2. **Identify the alternatives for achieving the objective.** Each alternative must satisfy the minimum performance requirements of the project.

3. **Establish the basic assumptions for the analysis.** These assumptions include specification of the base year for the analysis, the life-cycle study period, and the real discount rate.

4. **Identify, estimate, and determine the timing of all relevant costs.** Relevant costs are those costs that will be different among alternatives. Use the classification to be sure all costs are screened for inclusion. Be sure to consider all costs to direct users of the project, and any spillover costs associated with the project.

5. **Compute the life-cycle cost of each alternative** using the common data assumptions identified in step 3.

6. **Perform sensitivity analysis by re-computing the life-cycle cost for each alternative using different assumptions** about data inputs that are both relatively uncertain and significant in their impact on life-cycle cost.  Sensitivity analysis shows how sensitive a technology's costs are to uncertain data used in the economic analysis.

7. **Compare the alternatives' life-cycle costs** for each set of assumptions.

8. **Consider other project effects** — quantifiable and non-quantifiable — that are not included in the life-cycle cost calculations. If other effects are not equal and are considered significant, then turn to techniques such as multi-attribute decision analysis to account for all

types of benefits and costs.

9. **Select the best alternative.** Where other things are equal (e.g., performance and non-quantifiable impacts) select the economically efficient alternative with the minimum life-cycle cost, i.e., the greatest net savings compared to the base-case alternative.

## Requirements for an LCC Analysis

When using the LCC method, you must compute the life-cycle cost of two or more alternatives to measure cost effectiveness. The alternative with the minimum life-cycle cost is the most cost-effective option. If you make one of the alternatives a base case (usually the one with the lowest initial cost), you can compare the life-cycle cost of every other alternative against it to see which has the greatest net savings. The LCC and net savings approaches will both indicate the same best alternative.

Because we express future costs in our case study in constant or real dollars, we use a real discount rate. This means that you do not have to worry about inflation or deflation in arriving at your streams of future costs, because you are expressing costs in dollars of constant purchasing power, fixed on a calendar reference date, that exclude inflation or deflation (if your costs include inflation, however, you need to remove this inflation prior to using them in BridgeLCC). The real discount rate adjusts costs for the real earning opportunities of money over time. Government agencies tend to use real discount rates and constant dollars in their analyses.

Use the same fixed discount rate for all alternatives in a life-cycle cost comparison. Public projects typically are mandated to use a specific rate. Note that the economic viability of projects that save benefits or costs over time are very sensitive to the value of the discount rate. Figure A1 shows two significant effects that the discount rate has on present values of costs spread over time.

First, the present value of a given future cost amount decreases as the discount rate increases. For example, the present value of $1,000 ten years into the future drops from $613.91 at a discount rate of 5% (Point A) to $161.51 at a discount rate of 20% (Point B). Thus projects with cost savings spread into the future will generate larger present value net savings when evaluated with low rather than high discount rates.

Second, at any given discount rate, the farther into the future that any given amount occurs, the smaller its present value will be. Looking at the 5% discount rate line in Figure A1, $1,000 ten years out, worth $619.91 in present value (Point A), drops to a present value of $482.02 by year 15 (Point C).

Use the same study period for each alternative. The study period is the time over which the alternatives are compared. Using different study periods for different alternatives distorts the life-cycle cost measure. If project alternatives have different lives, include replacements in short-lived projects and consider the salvage value of long-lived projects to arrive at a common study period.

Implicit in any life-cycle cost analysis is the assumption that every proposed alternative will satisfy the minimum performance requirements of the project. These requirements include structural, safety, reliability, environmental, and specific building code requirements. Exclude from life-cycle cost analysis any alternatives that fail to meet the performance specifications of the project. If an alternative satisfies performance requirements and has additional positive features that are not explicitly accounted for in the life-cycle cost analysis, then consider an alternative economic measure such as net benefits.

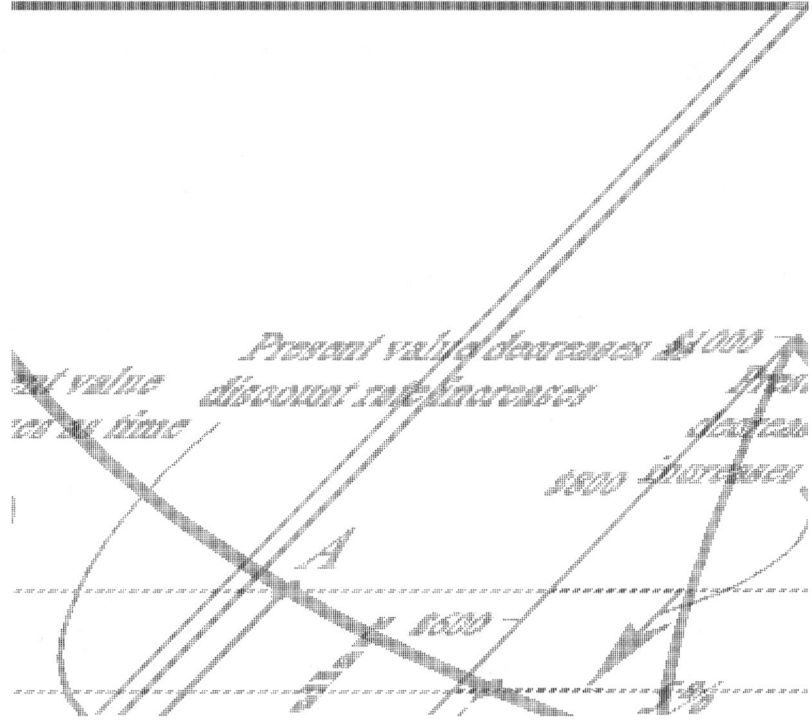

**Figure A1. Present Value of Future Costs, by Discount Rate**

## Applications of LCC

The LCC method has multiple applications in project evaluation. We look at each in turn as it applies to construction.

### Accept/Reject Decision
Choosing whether or not to do a project is an accept/reject decision. One example is deciding whether to coat an existing concrete bridge deck with polymer concrete asphalt or leave the deck "as is." The decision rule is to choose the alternative with minimum life-cycle cost.

### Material/Design Decision
A material/design decision occurs when you must choose the most cost effective of multiple material/design alternatives to satisfy an objective. The decision rule is to choose the material/design with minimum life-cycle cost. For example, given a particular material, what fabrication and construction method minimizes life-cycle cost? In this application, the decision has already been made to replace the deck with a particular material; the life-cycle cost analysis is needed to decide which design is most cost effective.

### Efficiency Level or Size Decision
Choosing how much of something to invest in is the efficiency level or size decision. An example is choosing the thickness of polymer-concrete asphalt to apply to a bridge deck. The decision rule is to choose the thickness of the coating that minimizes the life-cycle cost of the polymer-concrete road surface (where all thicknesses considered meet minimum performance requirements).

79

## A.2 The Cost Classification Scheme

There are two primary reasons for establishing a life-cycle cost classification or taxonomy when evaluating new-technology materials. First, the classification insures that all costs associated with the project are taken into account, and that these costs are accounted for in each alternative. This includes costs incurred by the owner/operator (agency costs), by direct users of the structure (user costs), and by organizations or individuals indirectly affected by the structure (spillover or third-party costs). Included are costs relating to the introduction of new materials (new-technology introduction [NTI] costs).

Second, the classification scheme allows for a detailed, consistent breakdown of the life-cycle cost and net savings estimates at several levels so that a clear picture can be had of the respective cost differences between material/design alternatives.

The classification scheme produces additional benefits such as providing a format for defining, collecting, and analyzing historical data for future projects; ensuring consistency in the data for economic evaluation of projects over time and from project to project; providing a check list for value engineering procedures; and providing a database format for computer-automated cost estimating.

The specifications of the classification scheme are general enough to cover the spectrum from privately owned and operated projects to publicly owned and operated projects.

The owners of some privately owned and operated structures might not include in their life-cycle cost analysis all of the user costs and spillovers that result from their projects; public agencies do not always incorporate such costs either. But environmental laws, for example, have forced private firms to internalize many spillover costs. And public agencies are beginning to treat user costs and other spillover costs as integral parts of their economic evaluations. Since new-technology materials are expected to have a significant impact on user costs, and public agencies are paying increasing attention to user costs in economic evaluations, it is important to include these costs in any life-cycle cost comparison of alternative materials.

### Costs by the Entity that Bears the Cost (Level 1)

### Agency Costs
Agency costs are all costs incurred by the project's owner or agent over the study period. These include but are not limited to design costs, capital costs, insurance, utilities, and servicing and repair of the facility. Agency costs are relatively easy to estimate for conventional material/designs since historical data on similar projects reveal these costs.

### User Costs
User costs accrue to the direct users of the project. For example, highway construction often causes congestion and long delays for private and commercial traffic. New bridge construction impacts traffic on the highway over which it passes. Maintenance and repair of an existing bridge, along with the rerouting of traffic, can impact drivers' personal time, as well as the operating cost of vehicles sitting in traffic. Accidents, involving harm to both vehicles and human life, tend to increase in road work areas.

These traffic delay costs, vehicle operating costs, and accident costs can be computed using simple formulas and tabulated traffic statistics from state departments of transportation. Similar types of user costs can be computed for projects where changes to buildings or other structures directly impact occupants.

**Third-Party Costs**

Third-party or spillover costs are all costs incurred by entities who are neither the agency/owners themselves nor direct users of the project. One example is the lost sales for a business establishment whose customer access has been impeded by construction of the project, or whose business property has been lost through the exercise of eminent domain. A second example is cost to humans and the environment from a construction process that pollutes the water, land, or atmosphere.

**Costs by LCC Category (Level 2)**

Level 2 groups costs according to the life-cycle categories typically used in the LCC formula: construction; operation, maintenance, and repair (OM&R); and disposal.

**Costs by Elemental Breakdown (Level 3)**

The third level of classification organizes costs (1) by specific functional element of the structure or facility, (2) by activities not assignable to functional elements (e.g., overhead), and (3) by any activities associated with the introduction of a new-technology material. Parts (1) and (2) are the traditional "elements" in an elemental cost estimate. We add part (3) on new-technology introduction costs to measure the unique costs of using a new material. We call these three groups an elemental classification.

**Elemental Costs**

Elements are major components of the project's structure, and are sometimes referred to as component systems or assemblies. Major elements that are common to most buildings, for example, are the foundation, superstructure, exterior closure, roofing, and interior. Elements common to bridges are superstructure, substructure, and approach. Each element performs a given function regardless of the materials used, design specified, or method of construction employed.

Individual cost estimates at the elemental level (e.g., $/square meter to furnish and install a concrete deck) are most useful in the pre-design stage when a variety of material/design combinations are being considered. This is the stage at which large net savings can be achieved by making economically optimal material/design choices. Detailed cost estimates of each alternative at the pre-design stage may not be economically feasible; elemental-based estimates, on the other hand, can be done quickly and are generally accurate enough to guide material/design decisions. Note, however, that for new-technology material/designs, there will not always be sufficient data to do element-based estimates; detailed products-based estimates and crew studies may be necessary.

BridgeLCC includes the PONTIS 2.0 element structure, which divides a bridge into four elements. Table A1 lists the elements and the bridge components assigned to each element. Use Table A1 to assign your individual costs to the correct element.

81

## Table A1. FHWA CORE Bridge Elements

| Element | Includes | |
|---|---|---|
| **Deck** | Concrete (Bare) | Steel - Open Grid |
| | Concrete Unprotected with AC Overlay | Steel - Concrete Filled Grid |
| | Concrete Protected with AC Overlay | Steel - Corrugated/Orthotropic/Etc. |
| | Concrete Protected with Thin Overlay | Timber (Bare) |
| | Concrete Protected with Rigid Overlay | Timber Protected with AC Overlay |
| | Concrete Protected with Coated Bars | |
| | Concrete Protected with Cathodic System | |
| **Superstructure** | Closed Web/Box Girder | Timber Truss/Arch |
| | Open Girder/Beam | Arch |
| | Stringer (stringer-floor beam system) | Cable (not embedded in concrete) |
| | Thru Truss (Bottom Chord) | Floor Beam |
| | Thru Truss (Excluding Bottom Chord) | Pin & Hanger Assembly |
| | Deck Truss | |
| **Substructure** | Column or Pile Extension | Submerged Pile Cap/Footing |
| | Pier Wall | Submerged Pile |
| | Abutment | Cap |
| | | Culvert |
| **Other** | Strip Seal Expansion Joint | Elastomeric Bearing |
| | Pourable Joint Seal | Movable Bearing (roller, sliding, etc.) |
| | Compression Joint Seal | Enclosed/Concealed Bearing |
| | Assembly Joint/Seal (Modular) | Fixed Bearing |
| | Open Expansion Joint | Pot Bearing |
| | Approach Slab w/ or wo/AC Overlay | Disk Bearing |
| | Bridge Railing | |

## Non-Elemental Costs

Non-elemental costs are all costs that cannot be attributed to specific functional elements of the project. A common example of a non-elemental agency cost is overhead expenses; a non-elemental third-party cost could be spillover costs. Because elemental cost categories are useful for generating and updating historical unit cost measures, all project costs that are not truly elemental must be excluded from these historical statistics and put in the non-elemental group.

## New-Technology Introduction (NTI) Costs

The final category contains costs directly associated with using a new material. The costs are generated from activities that insure that the designer is satisfied with the material's performance and predicted service life. Said another way, the NTI costs cover the activities that bring the material from the research laboratory to full field implementation. Figure A2 illustrates typical activities that occur in the new-technology introduction phase.

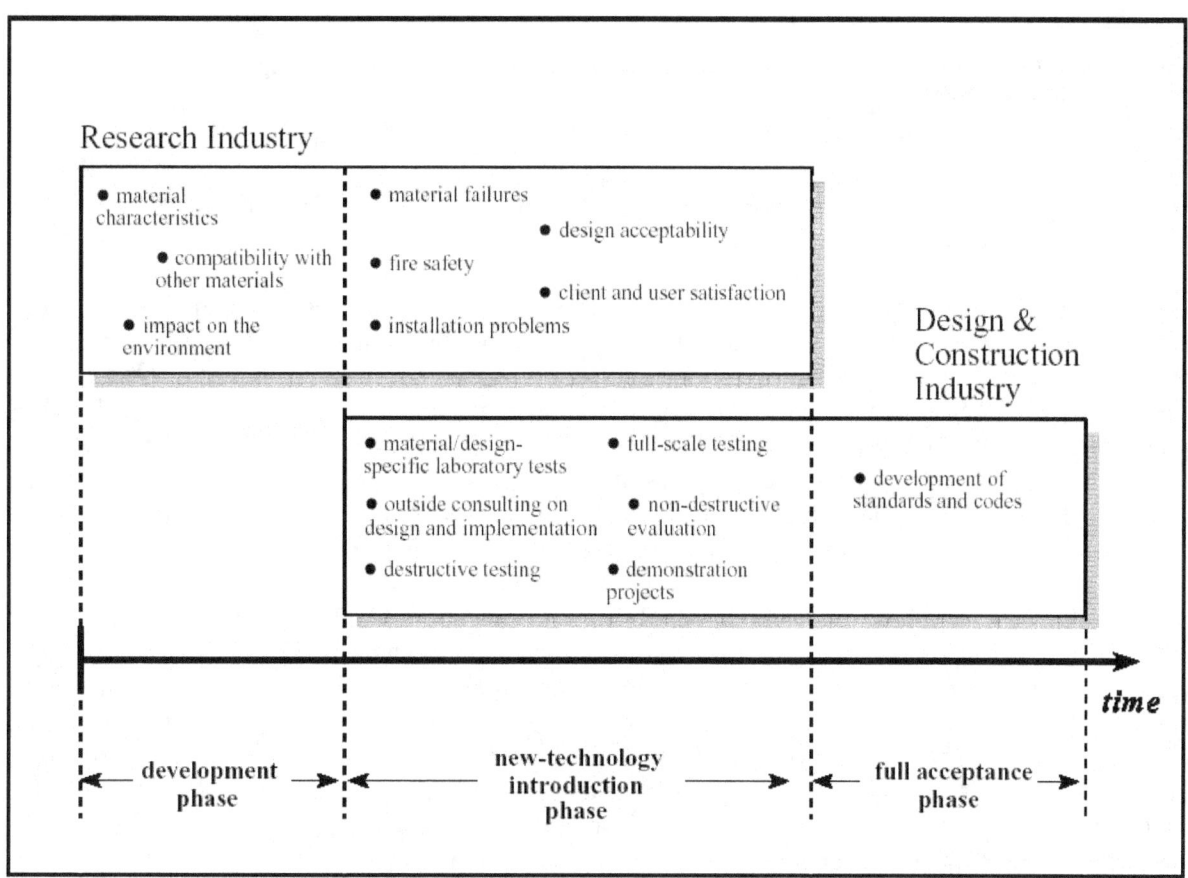

**Figure A2. Evolution of New Technology Materials**

In the development phase of a new material, laboratory researchers develop an understanding of the material's properties such as its structural and corrosive behavior and corrosion resistance, and how well it performs in conjunction with other materials. If promising applications are identified, both the research and construction industries will conduct activities that introduce and integrate the new-technology material to mainstream construction. These activities will include investigating material failures and installation problems and carrying out demonstration projects and non-destructive evaluation. If the material reaches full acceptance, these activities tend to diminish or stop.

New-technology introduction costs are all project-assignable costs. They include the extra time and labor to design, test, monitor, and use the new technology. These activities and costs disappear once the designer is satisfied with the technology's performance and service life, the technology enters full implementation, and its application has become routine. Examples of activities which help insure acceptability of a new-technology material and design include

> Full-scale testing and other laboratory tests;
> Demonstration projects;
> Hiring consultants and/or research institutions to assist in the evaluation process;
> The training of inspection, maintenance, and repair crews in the use of the new material;
> Non-destructive monitoring and evaluation of the new structure; and
> Additional material testing for government acceptance.

The costs of these activities can be directly estimated, as we do in the case study in Chapter 6.

**An Example of the Cost Classification Scheme**

As an example of how the cost classification is used to organize a life-cycle cost estimate, Figure A3 shows a schematic of a typical engineer's estimate.

Prior to public bidding of a highway overpass project, a state engineer estimates new construction costs by making a detailed quantity take-off of materials, and then assigning unit costs which reflect the labor, material, and equipment necessary to put the sub-component materials in place. These quantity take-offs are often structured by bridge component (level 3 project elements): bridge deck (element 1), substructure (element 2), and approach roadways (element 3). Non-elemental costs and new-technology introduction costs are then estimated and grouped as separate categories of level 3 costs. Next, because these level 3 elemental costs occur during initial construction, they are classified as level 2 initial construction costs. Finally, these are level 1 agency costs.

There are at least three benefits to this life-cycle cost classification of an engineer's estimate. First, it requires little to no restructuring of how current estimates are organized. Second, it insures proper identification and placement of costs due to its top-down and bottom-up functionality. The classification insures proper identification of all construction costs by allowing the estimator to start at the top of the classification (level 1) and work his or her way down each level. The classification's bottom-up ability is equally important: any estimate of a cost can be placed properly in the life-cycle cost classification by noting which entity bears the cost (level 1), which period in the life cycle the cost occurs (level 2), and what component of the project generates the cost (level 3).

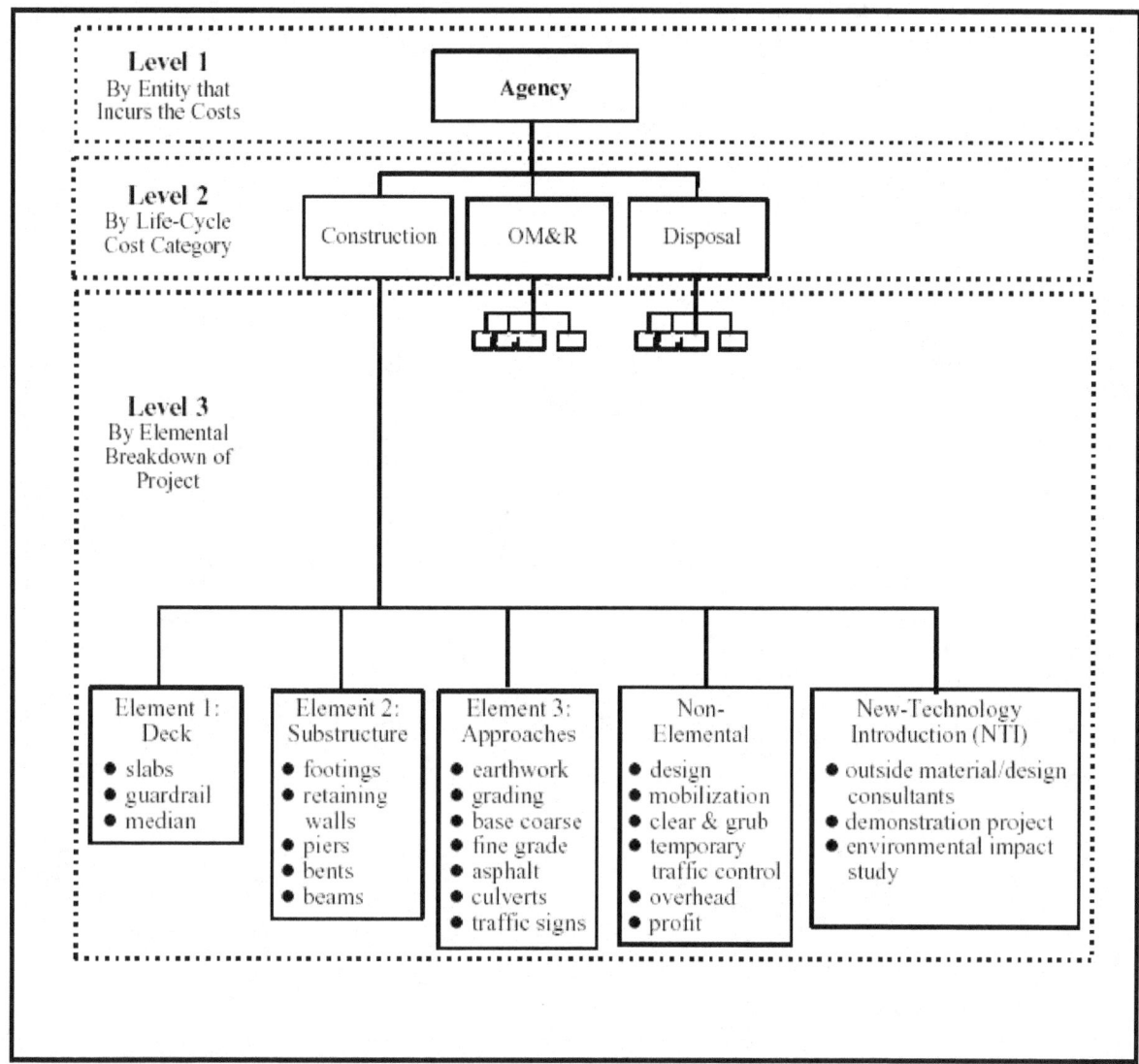

**Figure A3. An Example of the Cost Classification for an Engineer's Estimate of New Bridge Construction (with NTI Costs)**

The third benefit of this life-cycle cost classification is that actual construction costs classified by the same structural elements can be used to compile historical unit cost data on level 3 bridge element costs to be used in future life-cycle cost analyses.

# Appendix B. Life-Cycle Costing Formulas

## B.1  Basic Formula

Equation B1 shows the formula used in BridgeLCC to convert future costs to present value and sum them into a single life-cycle cost number.

$$PVLCC = \sum_{t=0}^{T} \frac{C_t}{(1 + d)^t},$$

(B1)

where

$C_t$ = the sum of all costs incurred at time $t$, valued in base-year dollars
$d$ = the real discount rate for converting time $t$ costs to time 0, and
$T$ = the number of time periods in the study period.

The unit of time used is typically the year; thus $C_t$ is the sum of all costs that occur in year $t$, and $T$ is the number of years in the study period.

## B.2  User Costs

User costs accrue to the direct users of the project. For example, highway construction often causes congestion and long delays for private and commercial traffic. New bridge construction impacts traffic on the highway over which it passes. Maintenance and repair of an existing bridge, along with the rerouting of traffic, can impact drivers' personal time and the operating cost of vehicles sitting in traffic. Accidents, involving harm to both vehicles and human life, tend to increase in road work areas.

These traffic delay costs, idle-capital costs, and accident costs can be computed using simple formulas and tabulated traffic statistics from state DOTs. BridgeLCC computes three types of user cost:

driver delay costs - the personal cost to drivers delayed by roadwork;
vehicle operating costs - the capital costs of vehicles delayed by roadwork; and
accident costs - the cost of damage to vehicles and injury to humans due to roadwork.

Equation B2 can be used to compute the cost to drivers of roadwork-related traffic delays.

$$\text{Driver Delay Costs} = \left( \frac{L}{S_a} - \frac{L}{S_n} \right) \times ADT \times N \times w,$$

(B2)

where

$L$ is the length of affected roadway or which cars drive,
$S_a$ is the traffic speed during bridge work activity,
$S_n$ is the normal traffic speed,
$ADT$ is the average daily traffic, measured in number of cars per day,

87

$N$ is the number of days of road work, and
$w$ is the dollar value of each hour of a driver's time.

The hourly value w is a weighted average of commercial vehicle drivers' and personal automobile drivers' time. Vehicle operating costs can be calculated using Equation B3.

$$\text{Vehicle Operating Costs} = \left( \frac{L}{S_a} - \frac{L}{S_n} \right) \times ADT \times N \times r, \quad \text{(B3)}$$

where r is a weighted-average vehicle cost similar to the weighted cost in Equation B2 and the remaining parameters are the same as those in Equation B2. Accident costs can be calculated using Equation B4

$$\text{Accident Costs} = L \times ADT \times N \times (A_a - A_n) \times c_a, \quad \text{(B4)}$$

where $c_a$ is the cost per accident, $A_a$ and $A_n$ are the during-construction and normal accident rates per vehicle-kilometer, and the remaining parameters are the same as those listed in Equations B2 and B3.

# References

American Society for Testing and Materials (ASTM). *ASTM Standards on Building Economics*, Fourth Edition (Philadelphia, PA: ASTM, 2001).

Bentz, Dale P., Clifton, James R., and Snyder, Kenneth A. "Predicting Service Life of Chloride-Exposed Steel-Reinforced Concrete," *Concrete International*, December 1996.

CALTRANS, DOTP - Transportation Economics. Life-Cycle Benefit/Cost Analysis (Highway Projects), 1995.

Ehlen, Mark A. *BridgeLCC 1.0 Users Manual*. NISTIR 6298. Gaithersburg, MD: National Institute of Standards and Technology, 1999.

Ehlen, Mark A., and Marshall, Harold E. *The Economics of New-Technology Materials: A Case Study of FRP Bridge Decking*. NISTIR 5864. Gaithersburg, MD: National Institute of Standards and Technology, 1996.

Norris, Gregory, and Marshall, Harold E. *Multi-attribute Decision Analysis Method for Evaluating Buildings and Building Systems*. NISTIR 5663. Gaithersburg, MD: National Institute of Standards and Technology, 1995.

Rosen, H. J., and Bennett, P. M. *Construction Materials Evaluation and Selection: A Systematic Approach* (New York, NY: John Wiley & Sons, Inc., 1979).

Ruegg, Rosalie T., and Marshall, Harold E. *Building Economics: Theory and Practice* (New York, N.Y.: Chapman & Hall, 1990).

Rushing, Amy S., and Fuller, Sieglinde K. *Energy Price Indices and Discount Factors for Life-Cycle Cost Analysis*. NISTIR 85-3273-18. Gaithersburg, MD: National Institute of Standards and Technology, 2003.